791.4302
L26g

Langman, Larry R.
 A guide to American film directors : the
sound era: 1929-1979 / Larry Langman. --
Metuchen, N.J. : Scarecrow, c1981.
 2v. ; 22 cm.
 Vol. 2: Film index.

 1. Moving-picture producers and directors--
United States. 2. Moving-pictures--United
States--Catalogs. I. Title.

7037102

A GUIDE TO AMERICAN FILM DIRECTORS

the sound era: 1929-1979

volume I

LARRY LANGMAN

The Scarecrow Press, Inc.
Metuchen, N.J., & London
1981

ALLEN COUNTY PUBLIC LIBRARY
FORT WAYNE, INDIANA

Library of Congress Cataloging in Publication Data

Langman, Larry.
  A guide to American film directors.

  Vol. 2: Film index.
  1. Moving-picture producers and directors--United States. 2. Moving-pictures--United States--Catalogs.
  I. Title.
  PN1998.A2L34   791.43'0233'0922   81-14536
  ISBN 0-8108-1467-6   AACR2

Copyright © 1981 by Larry Langman
Manufactured in the United States of America

PREFACE

The purpose of this volume is to present a list of American film directors and their complete works from 1929 to 1979. To keep the book manageable, silent films and television features were not considered. Otherwise, every known director of feature-length films has been included. Listed here are foreign directors who may have made at least one American film; filmmakers of hour-long documentaries; and American directors whose careers are almost entirely devoted to foreign films. Also included are the sound contributions of silent-film directors. Above all, every attempt was made to locate and verify all the titles and film dates for each director.

One difficulty concerned the transitional years, especially the films of 1929. Although some are all-talking, the degree to which sound is used in others makes them difficult to categorize. It was decided to include all of these movies, thus making the book more comprehensive and less puzzling to readers who may discover one of their favorite films missing.

The title of the book posed a second problem. Some directors, like Zeffirelli and Antonioni, clearly are not American, but some of their films (The Champ and Zabriskie Point, respectively) are American productions. Generally only these films are included for these directors, whereas the entire body of work of American directors in always listed. In any case the complete works of foreign directors are available in other easily attainable sources.

It is hoped that this guide will serve several purposes: first, that it will provide in one convenient volume a list of all American directors as well as all their films made during the first fifty years of the sound era; second, that it will aid in the rediscovery of the many early directors whose films are either lost or unknown to the general public; third, that this book will recognize the creativity of those young directors who will in time establish themselves as the next generation of major filmmakers.

The following sources have been invaluable in the compilation of this book and are suggested as additional references for the reader:

Blum, Daniel, ed. <u>Screen World</u> (annual). New York: Greenberg.
Catalog of Copyright Entries: Cumulative Series: <u>Motion Pictures</u>. Washington, D.C.: Copyright Office, Library of Congress, 1951.
<u>Film Daily Year Book of Motion Pictures</u> (1930 annually through 1969). New York: Distributed by Arno Press, 1970.
Geduld, Harry M. <u>The Birth of the Talkies</u>. Bloomington: Indiana University Press, 1975.
McCarthy, Todd, and Charles Flynn, eds. <u>Kings of the Bs</u>. New York: E. P. Dutton & Co., 1975.
Munden, Kenneth, ed. <u>American Film Institute Catalogue of Motion Pictures in the United States</u>. New York: R. R. Bowker Company, 1971.
Parish, James Robert, and Michael R. Pitts. <u>Film Directors</u>. Metuchen, N.J.: Scarecrow Press, 1974.
Willis, John, ed. <u>Screen World</u> (1930 annually through 1980). New York: Crown Publishers.

This book was planned as a complete, yet convenient one-volume guide to American film directors. However, to maximize its usefulness a companion volume is available. <u>A Guide to American Film Directors: Volume II: Title Index</u> lists every sound film alphabetically, with each entry containing release date and director. Together, they offer for the first time a practical and unique resource tool for the film scholar as well as the film buff.

ACKNOWLEDGMENTS

I would like to thank Sheila Lipstein for her clerical assistance. Also, I am gratefully indebted to the noted film historian William K. Everson for his many suggestions, revisions, and contributions.

ABBREVIATIONS USED IN THIS BOOK

    alt.   alternate title
    anim.   animation
    doc.   documentary
    orig.   original title
    ret.   retitled
    unc.   uncredited
    w/   with (co-director)

A
GUIDE
TO
AMERICAN
FILM
DIRECTORS

PAUL AARON

1978: A Different Story
1979: A Force of One

CHARLES ABBOTT

1937: The Fighting Texan
1939: The Adventures of the Masked Phantom

GEORGE ABBOTT (1889-)

1929: Half Way to Heaven; Why Bring That Up?
1930: Manslaughter; Sea God
1931: My Sin; The Cheat; Secrets of a Secretary; Stolen Heaven
1940: Too Many Girls
1957: The Pajama Game (w/ Stanley Donen)
1958: Damn Yankees (w/ Stanley Donen)

NORMAN ABBOTT

1966: The Last of the Secret Agents?

ALAN & JEANNE ABEL

1971: Is There Sex After Death?
1976: The Faking of the President

ROBERT ABEL

1972: Elvis on Tour (doc. w/ P. Adidge)
1973: Let the Good Times Roll (doc. w/ Sid Levin)

DENIS ABEY

1976: Never Too Young to Rock (doc.)

DERWIN ABRAHAMS

1941: Border Vigilantes; Secrets of the Vigilantes; Secrets of the Wasteland
1944: Cattle Call; Phantom Outlaws; Renegade Roundup; Return of the Durango Kid; Rough Riders Justice; Texas Rifles
1945: Both Barrels Blazing; Northwest Trail; Rustlers of the Badlands; Drifting Along; The Haunted Mine
1946: Fighting Frontiersman; Frontier Gunlaw; South of the Chisholm Trail
1947: Prairie Raiders; Riders of the Lone Star; The Stranger from Ponca City; Smoky River Serenade
1948: Docks of New Orleans; Cowboy Cavalier; The Rangers Ride; Steamboat Rhythm
1949: Mississippi Rhythm
1950: The Girl from San Lorenzo
1951: The Whistling Hills

(Serials)

1946: Chick Carter, Detective; Hop Harrigan; Son of the Guardsman
1948: Tex Granger
1953: Great Adventures of Capt. Kidd (w/ Charles Gould)

AL ADAMSON

1964: Two Tickets to Terror
1969: The Fakers; Gun Riders; Blood of Dracula's Castle
1970: Satan's Sadists; Five Bloody Graves; Hell's Bloody Devils
1971: Dracula vs. Frankenstein; Horror of the Blood Monsters; The Female Bunch; Last of the Comancheros
1972: Doomsday Voyage; Blood of Ghastly Horror; The Brain of Blood
1974: The Dynamite Brothers; Girls for Rent
1975: The Naughty Stewardesses; Stud Brown; Blazing Stewardesses
1976: Jessie's Girls; Black Heat

VICTOR ADAMSON (1890-1972)

1930: Sagebrush Politics; Sweeping Against the Winds; Desert Vultures
1931: Fighting Romance; Lightning Bill; Rawhide Romance; Ridin' Speed
1932: Boss Cowboy; Circle Canyon; Lightning Range; Range Busters
1933: The Fighting Cowboy
1934: Arizona Trail
1936: Desert Mesa
1938: Mormon Conquest
1963: Halfway to Hell

JUS ADDISS

1958: The Cry Baby Killer

PIERRE ADIDGE

1971: Joe Cocker (doc.)
1972: Elvis on Tour (doc. w/ Robert Abel)

JOSEPH ADLER

1971: Revenge Is My Destiny

LOU ADLER

1978: Up in Smoke

ED ADLUM

1976: Invasion of the Blood Farmers

JOHN G. ADOLFI (1888-1933)

1929: Evidence; Fancy Baggage; In the Headlines; Show of Shows
1930: College Lovers; Dumbbells in Ermine; Recaptured Love; Sinners' Holiday
1931: Alexander Hamilton; Compromised; The Millionaire
1932: A Successful Calamity; Central Park; The Man Who Played God
1933: The King's Vacation; The Working Man; Voltaire

FRANKLIN ADREON (1902-)

1955: No Man's Woman
1956: The Man Is Armed; Terror at Midnight
1957: Hell's Crossroads
1962: The Nun and the Sergeant
1966: Cyborg 2087; Dimension Five

(Serials)

1953: Canadian Mounties vs. Atomic Invaders
1954: Man with the Steel Whip; Trader Tom of the China Seas
1955: King of the Carnival; Panther Girl of the Jungle

GERARD ALCAN

1975: The Second Gun (doc.)

ROBERT ALDRICH (1918-)

1953: The Big Leaguer
1954: Apache; Vera Cruz; World for Ransom
1955: The Big Knife; Kiss Me Deadly
1956: Attack!; Autumn Leaves
1957: The Garment Jungle (unc. w/ V. Sherman)
1959: The Angry Hills; Ten Seconds to Hell
1961: The Last Sunset
1962: What Ever Happened to Baby Jane?; Sodom and Gomorrah (Brit. w/ S. Leone)
1963: Four for Texas
1965: Hush ... Hush, Sweet Charlotte
1966: Flight of the Phoenix
1967: The Dirty Dozen
1968: The Killing of Sister George; The Legend of Lylah Clare
1970: Too Late the Hero
1971: The Grissom Gang
1972: Ulzana's Raid
1973: Emperor of the North Pole (ret. Emperor of the North)
1974: The Longest Yard
1975: Hustle
1977: The Choirboys; Twilight's Last Gleaming
1979: The Frisco Kid

JAMES ALGAR (1912-)

1949: The Adventures of Ichabod and Mr. Toad
1953: The Living Desert (doc.)
1954: The Vanishing Prairie (doc.)
1955: The African Lion (doc.)
1956: Secrets of Life (doc.)
1958: White Wilderness (doc.)
1960: Jungle Cat (doc.)
1962: The Legend of Lobo

SIDNEY ALGIER

1931: Wild Horse (w/ R. Thorpe)

HOWARD ALK

1975: Janis (doc. w/ S. Findlay)

WILLIAM ALLAND (1916-)

1916: Look in Any Window

CORY ALLEN (1934-)

1971: Pinocchio
1977: Thunder and Lightning
1978: Avalanche

FRED ALLEN

1931: Freighters of Destiny
1932: Ride Him Cowboy
1933: The Mysterious Rider; Beyond the Rockies; Ghost Valley; Partners; Saddle Buster

IRVING ALLEN (1905-)

1946: Avalanche; High Conquest; Strange Voyage
1948: 16 Fathoms Deep
1951: Slaughter Trail

IRWIN ALLEN (1916-)

1956: The Animal World
1957: The Story of Mankind
1960: The Lost World
1961: Voyage to the Bottom of the Sea
1962: Five Weeks in a Balloon
1974: The Towering Inferno (w/ John Guillermin; Allen directed fire sequences)
1978: The Swarm
1979: Beyond the Poseidon Adventure

LEWIS ALLEN (1905-)

1944: Our Hearts Were Young and Gay; The Uninvited
1945: The Unseen; Those Endearing Young Charms
1946: Desert Fury; The Imperfect Lady; The Perfect Marriage
1948: Sealed Verdict; So Evil My Love
1949: Chicago Deadline
1951: Appointment with Danger; Valentino
1952: At Sword's Point
1954: Suddenly
1955: A Bullet for Joey; Illegal
1958: Another Time, Another Place
1959: Whirlpool (Brit.)

WOODY ALLEN (1935-)

- 1966: What's Up, Tiger Lily?
- 1969: Take the Money and Run
- 1971: Bananas
- 1972: Everything You Always Wanted to Know About Sex (but Were Afraid to Ask)
- 1973: Sleeper
- 1975: Love and Death
- 1977: Annie Hall
- 1978: Interiors
- 1979: Manhattan

JOHN A. ALONZO

- 1978: FM

HERBERT S. ALTMAN

- 1971: Dirtymouth

ROBERT ALTMAN (1925-)

- 1957: The Delinquents; The James Dean Story (doc. w/ G. W. George)
- 1968: Countdown
- 1969: That Cold Day in the Park
- 1970: Brewster McCloud; M*A*S*H
- 1971: McCabe and Mrs. Miller
- 1972: Images
- 1973: The Long Goodbye
- 1974: California Split; Thieves Like Us
- 1975: Nashville
- 1976: Buffalo Bill and the Indians
- 1977: Three Women
- 1978: A Wedding
- 1979: A Perfect Couple; Quintet

ROBERT ALTON (1906-1957)

- 1947: Merton of the Movies
- 1950: Pagan Love Song

ROD (RODNEY) AMATEAU (1923-)

- 1952: The Bushwhackers
- 1953: Monsoon
- 1969: Pussycat, Pussycat, I Love You

1971: The Statue
1972: Where Does It Hurt?
1976: Drive-In

ROBERT AMRAM

1975: Pacific Challenge

GEORGE AMY (1903-)

1933: She Had to Say Yes (w/ B. Berkeley)
1939: Kid Nightingale
1940: Gambling on the High Seas; Granny Get Your Gun

JOHN MURRAY ANDERSON

1930: King of Jazz

MARION CLAYON ANDERSON

1960: The Crowning Experience (doc.)

MICHAEL ANDERSON (1920-) Brit.

(American films only)
1956: Around the World in 80 Days
1959: The Wreck of the Mary Deare
1960: All the Fine Young Cannibals
1963: Flight From Ashiya (U.S./Jap.)
1964: Wild and Wonderful.
1968: The Shoes of the Fisherman
1975: Doc Savage: The Man of Bronze
1976: Logan's Run
1977: Orca

ROBERT ANDERSON

1975: Swiss Bank Account

DEL ANDREWS

1929: The Wild West Show

EDMOND ANGELO

1952: Breakdown

KEN ANNAKIN (1914-   ) Brit.

(American films only)
1959: Third Man on the Mountain
1960: Swiss Family Robinson
1962: The Longest Day (w/ A. Marton, B. Wicki)
1965: Battle of the Bulge
1968: The Biggest Bundle of Them All
1969: Those Daring Young Men in Their Jaunty Jalopies (Brit./U.S.)
1976: Paper Tiger

JOSEPH ANTHONY (1912-   )

1956: The Rainmaker
1958: The Matchmaker
1959: Career
1961: All in a Night's Work
1965: Conquered City (Ital.)
1972: Tomorrow

MICHELANGELO ANTONIONI (1912-   ) Ital.

(American films only)
1970: Zabriskie Point

MICHAEL APTED (Brit.)

(American films only)
1979: Agatha

MANUEL ARANGO

1974: Sky High (doc.)

GEORGE ARCHAINBAUD (1890-1959)

1929: Broadway Scandals; College Coquette; George Washington Cohen; Man in Hobbles; Two Men and a Maid; Voice Within
1930: Alias French Gertie; Framed; The Broadway Hoofer; The Silver Horde; Shooting Straight
1931: The Lady Refuses; Three Who Loved
1932: Men of Chance; State's Attorney; The Lost Squadron; The Penguin Pool Murder; Thirteen Women.
1933: After Tonight; The Big Brain
1934: Keep 'Em Rolling; Murder on the Blackboard
1935: My Marriage; Thunder in the Night
1936: The Return of Sophie Lang

1937: Blonde Trouble; Clarence; Hideaway Girl; Hotel Haywire; Thrill of a Lifetime
1938: Boy Trouble; Campus Confessions; Her Jungle Love; Thanks for the Memory
1939: Night Work; Some Like It Hot
1940: Comin' Round the Mountain; Opened by Mistake; Untamed
1942: Flying with Music
1943: False Colors; Hoppy Serves a Writ; The Kansan; The Woman of the Town
1944: Alaska; Mystery Man; Texas Masquerade
1945: Girls of the Big House; The Big Bonanza
1946: Fool's Gold; The Devil's Playground; Unexpected Guest
1947: Dangerous Venture; Hoppy's Holiday; King of the Wild Horses; The Marauders; The Millerson Case
1948: False Paradise; Silent Conflict; Strange Gamble; The Dead Don't Dream
1950: Border Treasure; Hunt the Man Down
1952: Apache Country; Barbed Wire; Blue Canadian Rockies; Night Stage to Galveston; The Old West; The Rough, Tough West; Wagon Team
1953: Goldtown Ghost Riders; On Top of Old Smoky; Pack Train; Last of the Pony Riders; Saginaw Trail; Winning of the West

ALAN ARKIN (1934-)

1971: Little Murders
1977: Fire Sale

ROBERT ARKLESS

1975: The Man Who Would Not Die

ALLAN ARKUSH

1978: Deathsport (w/ H. Suso)
1979: Rock 'n' Roll High School

GEORGE ARMITAGE

1972: Hit Man; Private Nurses
1976: Vigilante Force

JACK ARNOLD (1916-)

1953: Girls in the Night; It Came from Outer Space; The Glass Web
1954: Creature from the Black Lagoon
1955: Revenge of the Creature; Tarantula; The Man from Bitter Ridge

1956: Outside the Law; Red Sundown
1957: Man in the Shadow; The Incredible Shrinking Man; The Tattered Dress
1958: High School Confidential; Monster on the Campus; The Lady Takes a Flyer; The Space Children
1959: No Name on the Bullet; The Mouse That Roared (Brit.)
1961: Bachelor in Paradise
1964: A Global Affair; The Lively Set
1969: Hello Down There
1974: Black Eye
1975: Boss Nigger; The Swiss Conspiracy (Germ./U.S.)

NEWTON ARNOLD

1962: Hands of a Stranger

KAREN ARTHUR

1976: Legacy
1979: The Mafu Cage

DOROTHY ARZNER (1900-)

1929: The Wild Party
1930: Anybody's Woman; Paramount on Parade (w/ others); Sarah and Son
1931: Honor Among Lovers; Working Girls
1932: Merrily We Go to Hell
1933: Christopher Strong
1934: Nana
1936: Craig's Wife
1937: The Bride Wore Red
1940: Dance, Girl, Dance
1943: First Comes Courage

STEVEN ASCHER

1978: Life and Other Anxieties (doc. w/ Ed Pincus)

DEREK ASHBURNE

1970: The Politicians

HAL ASHBY (1936-)

1970: The Landlord
1971: Harold and Maude

1973: The Last Detail
1975: Shampoo
1976: Bound for Glory
1978: Coming Home
1979: Being There

RICHARD ASHE

1976: Track of the Moon Beast

WILLIAM ASHER (c. 1919-)

1948: Leather Gloves (w/ Richard Quine)
1957: The Shadow on the Window; The 27th Day
1963: Beach Party; Johnny Cool
1964: Bikini Beach; Muscle Beach Party
1965: Beach Blanket Bingo; How to Stuff a Wild Bikini
1966: Fireball 500

RAY ASHLEY

1953: The Little Fugitive

THOMAS ATKINS

1934: The Silver Streak; Mutiny Ahead
1946: Hi Gaucho

RICHARD ATTENBOROUGH (1923-) Brit.

(American films only)
1978: Magic

JOHN H. AUER (1909-) Hung.

(American films only)
1935: The Crime of Dr. Crespi
1937: A Man Betrayed; Circus Girl; Rhythm in the Clouds
1938: A Desperate Adventure; Invisible Enemy; I Stand Accused; Orphans of the Street; Outside of Paradise
1939: Calling All Marines; Forged Passport; S.O.S. Tidal Wave; Smuggled Cargo
1940: The Hit Parade of 1941; Thou Shalt Not Kill; Women in War
1941: The Devil Pays Off
1942: Johnny Doughboy; Moonlight Masquerade; Pardon My Stripes
1943: Gangway for Tomorrow; Tahiti Honey
1944: Music in Manhattan; Seven Days Ashore

1945: Pan-Americana
1947: Beat the Band; The Flame
1948: I, Jane Doe; Angel on the Amazon
1950: Hit Parade of 1951; The Avengers
1952: Thunderbirds
1953: The City That Never Sleeps
1954: Hell's Half Acre
1955: The Eternal Sea
1957: Johnny Trouble

ROBERT ALAN AURTHUR (1922-1978)

1969: The Lost Man

ARAM AVAKIAN

1962: Lad: A Dog (w/ L. H. Martinson)
1970: The End of the Road
1973: Cops and Robbers
1974: 11 Harrowhouse (Brit.)

HIKMET AVEDIS

1973: The Stepmother
1974: The Teacher
1975: The Specialist; Dr. Minx
1976: Scorchy

HY AVERBACK (c.1925-)

1966: Chamber of Horrors
1968: I Love You, Alice B. Toklas!; Where Were You When the Lights Went Out?
1969: The Great Bank Robbery
1970: Suppose They Gave a War and Nobody Came?

JOHN G. AVILDSEN (1937-)

1969: Turn on to Love
1970: Joe
1971: Cry Uncle; Okay Bill
1973: Save the Tiger
1974: The Stoolie
1975: W. W. and the Dixie Dancekings; Foreplay (w/ others)
1976: Rocky
1978: Slow Dancing in the Big City

GEORGE AXELROD (1922-   )

1966: Lord Love a Duck
1968: The Secret Life of an American Wife

LEW AYRES (1908-   )

1936: Hearts in Bondage
1955: Altars of the East

LLOYD BACON (1890-1955)

1929: Honky Tonk; No Defense; Say It with Songs; Stark Mad
1930: A Notorious Affair; Moby Dick; She Couldn't Say No; So Long Lefty; The Office Wife; The Other Tomorrow
1931: Fifty Million Frenchmen; Gold Dust Gertie; Honor of the Family; Kept Husbands; Sit Tight
1932: Crooner; Fireman, Save My Child; Manhattan Parade; Miss Pinkerton; The Famous Ferguson Case; You Said a Mouthful
1933: 42nd Street; Mary Stevens, M.D.; Picture Snatcher; Son of a Sailor; Footlight Parade
1934: A Very Honorable Guy; Here Comes the Navy; He Was Her Man; Six-Day Bike Rider; Wonder Bar
1935: Broadway Gondolier; Devil Dogs of the Air; Frisco Kid; In Caliente; The Irish in Us
1936: Cain and Mabel; Gold Diggers of 1937; Sons o' Guns
1937: Ever Since Eve; Marked Woman; San Quentin; Submarine
1938: A Slight Case of Murder; Boy Meets Girl; Cowboy from Brooklyn; Racket Busters
1939: Espionage Agent; Indianapolis Speedway; The Oklahoma Kid; Wings of the Navy
1940: A Child Is Born; Brother Orchid; Knute Rockne--All-American; Invisible Stripes; Three Cheers for the Irish
1941: Affectionately Yours; Footsteps in the Dark; Honeymoon for Three; Navy Blues
1942: Larceny, Inc.; Silver Queen; Wings for the Eagle
1943: Action in the North Atlantic
1944: Sunday Dinner for a Soldier; The Sullivans
1945: Captain Eddie
1946: Home Sweet Homicide; Wake Up and Dream
1947: I Wonder Who's Kissing Her Now?
1948: Don't Trust Your Husband; Give My Regards to Broadway; You Were Meant for Me
1949: It Happens Every Spring; Miss Grant Takes Richmond; Mother Is a Freshman
1950: The Fuller Brush Girl; The Good Humor Man; Kill the Umpire
1951: Call Me Mister; The Golden Girl; The Frogmen

1953: The Great Sioux Uprising; The I Don't Care Girl; Walking My Baby Back Home
1954: She Couldn't Say No; The French Line

CLARENCE BADGER (1880-1964)

1929: Paris
1930: No, No, Nanette; Murder Will Out; Sweethearts and Wives; The Bad Man
1931: The Hot Heiress; Woman Hungry; Party Husband
1933: When Strangers Marry
1939: Rangle River

JOHN BADHAM

1976: The Bingo Long Traveling All Stars and the Motor Kings
1977: Saturday Night Fever
1979: Dracula (Brit.)

REZA S. BADIYL

1973: Trader Horn

MAX BAER

1975: The Wild McCullachs
1976: Ode to Billy Joe
1979: Hometown, U.S.A.

CHUCK BAIL

1974: Black Sampson
1975: Cleopatra Jones and the Casino of Gold
1976: The Gumball Rally

REX BAILEY

1953: Mexican Manhunt; Fangs of the Arctic; Northern Patrol

RICHARD BAILEY

1975: Win, Place or Steal

FRED BAKER

1970: Events

1975: Lenny Bruce Without Tears (doc.)

ROY BAKER (Brit.)
(American films only)

1951: I'll Never Forget You
1952: Night Without Sleep; Don't Bother to Knock
1953: Inferno

RALPH BAKSHI (1939-)

1972: Fritz the Cat (anim.)
1973: Heavy Traffic (anim.)
1975: Coonskin (anim.)
1977: Wizards (anim.)
1978: The Lord of the Rings (anim.)

BURT BALABAN (1922-1965)

1957: Lady of Vengeance (Brit.)
1958: High Hell
1960: Murder, Inc. (w/ S. Rosenberg)
1961: Mad Dog Coll
1966: The Gentle Rain

CARROLL BALLARD

1979: The Black Stallion

JOHN BALLARD

1979: The Orphan

ALBERT BAND (1924-)

1956: The Young Guns
1958: I Bury the Living
1959: Face of Fire
1966: The Tramplers (Ital.)
1978: Dracula's Dog
1979: She Came to the Valley

CHARLES BAND

1977: Crash!

MONTY BANKS (1897-1950) Brit.

(American films only)
1941: Great Guns

JACK BARAN

1971: Roommates

JOSEPH BARBERA (1911-)

1964: Hey There It's Yogi Bear (anim. w/ William Hanna)
1966: The Man Called Flintstone (anim. w/ William Hanna)

J. A. BARDEW

1969: The Last Day of the War

RICHARD L. BARE (1909-)

1948: Smart Girls Don't Talk
1949: Flaxy Martin; The House Across the Street
1950: Return of the Frontiersman; This Side of the Law
1953: Prisoners of the Casbah
1956: The Outlanders; The Storm Riders
1957: Shoot-Out at Medicine Bend; The Travellers
1958: Girl on the Run
1960: This Rebel Breed
1968: I Sailed to Tahiti with an All Girl Crew
1973: Wicked, Wicked

BRADLEY BARKER

1929: Mother's Boy

REGINALD BARKER (1886-1937)

1929: Seven Keys to Baldpate; New Orleans; The Mississippi Gambler
1930: The Great Divide; Hide Out
1934: The Moonstone
1935: The Healer; Women Must Dress
1936: Forbidden Heaven

WILLIAM H. BARNETT

1975: Nothing by Chance

JOHN BARNWELL

1956: Huk!
1959: Surrender--Hell!

ALLEN BARON

1966: Terror in the City
1972: Outside In

ARTHUR BARRON

1973: Jeremy
1977: Brothers

NICHOLAS BARROWS

1937: Dangerous Holiday

DONALD BARRY (1912-1980)

1954: Jesse James' Women

MICHAEL BARRY

1974: The Second Coming of Suzanne

WESLEY BARRY

1952: The Steel Fist
1954: The Outlaw's Daughter

DICK BARRYMORE

1969: Last of the Ski Bums

LIONEL BARRYMORE (1878-1954)

1929: His Glorious Night; Madame X; Unholy Night
1930: Rogue Song
1931: Ten Cents a Dance

LEON BARSHA

1937: One Man Justice; Trapped; Two Fisted Sheriff; Two Gun Law

1938: Who Killed Gail Preston?; Convicted
1939: Manhattan Shakedown; Special Inspector
1952: The Place That Thrills

PAUL BARTEL

1972: Private Parts
1975: Death Race 2000
1976: Cannonball

HALL BARTLETT (1922-)

1955: Unchained
1957: Drango (w/ Jules Bricken); Zero Hour
1960: All the Young Men
1963: The Caretakers
1969: Changes
1971: The Sandpit Generals (The Wild Pack)
1973: Jonathan Livingston Seagull

RICHARD BARTLETT

1955: The Lonesome Trail; The Silver Star
1956: I've Lived Before; Rock, Pretty Baby; Two Gun Lady
1957: Joe Dakota; Slim Carter
1958: Money, Women and Guns

CHARLES T. BARTON (1902-)

1934: Wagon Wheels
1935: Car No. 99; Rocky Mountain Mystery; The Last Outpost (w/ Louis J. Gasnier)
1936: And Sudden Death; Nevada; Murder with Pictures; Rose Bowl; Timothy's Quest
1937: Forlorn River; The Crime Nobody Saw; Thunder Train
1938: Born to the West
1939: Behind Prison Gates; Five Little Peppers and How They Grew
1940: Babies for Sale; Five Little Peppers at Home; Five Little Peppers in Trouble; Island of Doomed Men; My Son Is Guilty; Nobody's Children; Out West with the Peppers
1941: Harmon of Michigan; Honolulu Lu; Sing for Your Supper; Two Latins from Manhattan; The Big Boss; The Phantom Submarine; The Richest Man in Town
1942: A Man's World; Hello, Annapolis; Lucky Legs; Laugh Your Blues Away; Parachute Nurse; Shut My Big Mouth; Sweetheart of the Fleet; The Spirit of Stanford; Tramp, Tramp, Tramp
1943: Is Everybody Happy?; Let's Have Fun; Reveille with Beverly; She Has What It Takes; What's Buzzin', Cousin?

1944: Beautiful but Broke; Hey, Rookie; Jam Session; Louisiana Hayride
1945: Men in Her Diary; The Beautiful Cheat
1946: Smooth as Silk; The Time of Their Lives; White Tie and Tails; The Ghost Steps Out
1947: Buck Privates Come Home; The Wistful Widow of Wagon Gap
1948: Abbott and Costello Meet Frankenstein; Mexican Hayride; The Noose Hangs High
1949: Abbott and Costello Meet the Killer; Africa Screams; Free For All
1950: Double Crossbones; The Milkman
1952: Ma and Pa Kettle at the Fair
1956: Dance with Me, Henry
1959: The Shaggy Dog
1960: Toby Tyler
1962: Swingin' Along

DON BARTON

1975: Blood Waters of Dr. Z

JULES BASS

1966: The Daydreamer
1967: Mad Monster Party

SAUL BASS (1920-)

1974: Phase IV

RICK BAXTER

1975: Ali the Man: Ali the Fighter (doc. w/ William Greaves)

SCOTT R. BEAL

1935: Straight from the Heart
1938: Convicts at Large (w/ D. Friedman)

ROBERT B. BEAN

1971: Made for Each Other

WARREN BEATTY (1937-)

1978: Heaven Can Wait (w/ Buck Henry)

WILLIAM BEAUDINE (1890-1970)

1929: Fugitives; Hard to Get; The Girl from Woolworths; Two Weeks Off
1930: The Road to Paradise; Those Who Dance; Wedding Rings
1931: Misbehaving Ladies: Penrod and Sam; The Lady Who Dared; The Mad Parade; The Men in Her Life; Father's Son
1932: Make Me a Star!; Three Wise Girls
1933: Her Bodyguard; The Crime of the Century
1934: The Old-Fashioned Way
1935: Boys Will Be Boys (Brit.); Dandy Dick (Brit); Get Off My Foot (Brit.); Two Hearts in Harmony (Brit.)
1936: Educated Evans (Brit.); It's in the Bag (Brit.); Mr. Cohen Takes a Walk (Brit.); Where There's a Will (Brit.); Windbag the Sailor (Brit.)
1937: Feather Your Nest (Brit.); Transatlantic Trouble (Brit.)
1938: Said O'Reilly to McNab (Brit.); Torchy Gets Her Man
1939: Torchy Blane in Chinatown
1940: Misbehaving Husbands
1941: Desperate Cargo; Emergency Landing; Federal Fugitives; Mr. Celebrity; The Blonde Comet
1942: Duke of the Navy; Foreign Agent; Gallant Lady; Men of San Quentin; One Thrilling Night; Phantom Killer; The Broadway Big Shot; The Living Ghost; The Miracle Kid; The Panther's Claw; Professor Creeps; Prison Girls
1943: Clancy Street Boys; Ghosts on the Loose; Here Comes Kelly; Mr. Muggs Steps Out; Spotlight Scandals; The Ape Man; The Mystery of the 13th Guest
1944: Bowery Champs; Crazy Knights; Detective Kitty O'Day; Follow the Leader; Hot Rhythm; Leave It to the Irish; Mom and Dad; Oh, What a Night; Shadow of Suspicion; Voodoo Man; What a Man!
1945: Black Market Babies; Blonde Ransom; Come Out Fighting; Fashion Model; Swingin' on a Rainbow; The Adventures of Kitty O'Day
1946: Below the Deadline; Don't Gamble with Strangers; Girl on the Spot; Mr. Hex; One Exciting Week; Spook Busters
1947: Bowery Buckaroos; Gas House Kids Go West; Hard Boiled Mahoney; Killer at Large; News Hounds; Philo Vance Returns; The Chinese Ring (The Red Hornet); Too Many Winners
1948: Angels' Alley; Incident; Jiggs and Maggie in Court (w/ Edward F. Cline); Jinx Money; Kidnapped; Smuggler's Cove; The Feathered Serpent; The Golden Eye; The Shanghai Chest
1949: Forgotten Women; Jackpot Jitters; Tough Assignment; Tuna Clipper
1950: Again Pioneers! Blue Grass of Kentucky; Blues Busters; Blonde Dynamite; County Fair; Jiggs and Maggie Out West; Lucky Losers; Second Chance
1951: Bowery Battalion; Crazy over Horses; Cuban Fireball; Ghost Chasers; Havana Rose; Let's Go Navy; The Prince of Peace (w/ Harold Daniels)
1952: Bela Lugosi Meets a Brooklyn Gorilla; Feudin' Fools; Here

      Come the Marines; Hold That Line; Jet Job; No Holds
      Barred; Rodeo; The Rose Bowl Story
1953: For Every Child; Jalopy; Murder Without Tears; Roar of
      the Crowd; The Hidden Heart
1954: More for Peace; Paris Playboys; Pride of the Blue Grass;
      Yukon Vengeance
1955: Each According to His Faith; High Society; Jail Busters
1957: In the Money; Up in Smoke; Westward Ho the Wagons!
1960: Ten Who Dared
1963: Lassie's Great Adventure
1966: Billy the Kid vs. Dracula; Jesse James Meets Frankenstein's
      Daughter

HARRY BEAUMONT (1888-1966)

1929: The Broadway Melody
1930: The Floradora Girl; Our Blushing Brides; Lord Byron of
      Broadway (w/ W. Nigh); Those 3 French Girls; Children of
      Pleasure
1931: Dance, Fools, Dance; Laughing Sinners; The Great Lover
1932: Faithless; Unashamed; Are You Listening?; West of Broadway
1933: When Ladies Meet; Made on Broadway; Should Ladies Behave?
1934: Murder in the Private Car
1935: Enchanted April
1936: The Girl on the Front Page
1937: When's Your Birthday?
1944: Maisie Goes to Reno
1945: Twice Blessed
1946: Up Goes Maisie; The Show-Off
1947: Undercover Maisie
1948: Alias a Gentleman

GEORGE BECK

1951: Behave Yourself!

MARTIN BECK

1974: Challenge
1975: The Brass Ring

HAROLD BECKER

1979: The Onion Field

TERRY BECKER

1975: The Thirsty Dead

VERNON F. BECKER

1969: The Funniest Man in the World (doc.)

FORD BEEBE (1888-)

1932: The Pride of the Legion
1933: Laughing at Life
1935: Law Beyond the Range; The Man from Guntown
1936: Stampede
1937: Trouble at Midnight; Westbound Limited
1939: Oklahoma Frontier
1940: Son of Roaring Dan
1941: The Masked Rider
1942: The Night Monster
1943: Frontier Badmen
1944: Enter Arsene Lupin; The Invisible Man's Revenge
1945: Easy to Look At
1946: My Dog Shep
1947: Six Gun Serenade
1948: Courtin' Trouble; Shep Comes Home
1949: Bomba, the Jungle Boy; Bomba on Panther Island; Red Desert; Satan's Cradle; The Dalton Gang
1950: Bomba and the Hidden City; The Lost Volcano
1951: Elephant Stampede; The Lion Hunters
1952: African Treasure; Bomba and the Jungle Girl; Wagons West
1953: Safari Drums
1954: Killer Leopard; The Golden Idol
1956: Lord of the Jungle

(Serials)

1932: The Shadow of the Eagle (w/ B. Eason); The Last of the Mohicans (w/ B. Eason); The Pride of the Legion
1935: The Adventures of Rex and Rinty (w/ B. Eason)
1936: Ace Drummond (w/ Colbert Smith)
1937: Jungle Jim (w/ C. Smith); Secret Agent X-9 (w/ C. Smith); Wild West Days (w/ C. Smith); Radio Patrol (w/ C. Smith)
1938: Tim Tyler's Luck (w/ W. Gittens, C. Smith); Flash Gordon's Trip to Mars (w/ R. E. Hill); Red Barry (w/ A. James)
1939: Buck Rogers; The Phantom Creeps; The Oregon Trail; The Green Hornet (all w/ S. A. Goodkind)
1940: Flash Gordon Conquers the Universe (w/ R. Taylor); Winners of the West (w/ R. Taylor; Junior G-Men (w/ J. Rawlins)

1941: Sky Raiders; Riders of Death Valley; Don Winslow of the Navy (all w/ R. Taylor); Sea Raiders (w/ J. Rawlins)
1942: Overland Mail (w/ J. Rawlins)

JACQUE BEERSON

1970: The Dark Side of Tomorrow (w/ Barbara Peeters)

MONTA BELL (1891-1958)

1930: East Is West; Young Man of Manhattan
1931: Fires of Youth; Personal Maid; Up for Murder
1932: Downstairs
1933: The Worst Woman in Paris?

EARL BELLAMY (1917-)

1956: Blackjack Ketchum, Desperado
1958: Toughest Gun in Tombstone
1962: Stagecoach to Dancer's Rock
1965: Fluffy
1966: Gunpoint; Incident at Phantom Hill; Munster, Go Home!
1968: Three Guns for Texas (w/ David Lowell Rich and Paul Stanley)
1969: Backtrack
1974: Sidecar Boys (Australia)
1975: Seven Alone; Sidecar Racers; Part 2 Walking Tall; Against a Crooked Sky
1977: Sidewinder 1; Speedtrap

JOEL BENDER

1979: Gas Pump Girls

LASLO BENEDEK (1907-)

1948: The Kissing Bandit
1949: Port of New York
1951: Death of a Salesman
1952: Storm over the Tiber
1953: The Wild One
1954: Bengal Brigade

1957: Affair in Havana
1960: Malaga (Brit.)
1966: Namu, the Killer Whale
1968: Daring Game
1970: The Night Visitor
1976: Assault on Agathon

RICHARD BENEDICT (1923-)

1965: Winter a Go-Go
1969: Impasse

TONY BENEDICT

1970: Santa and the Three Bears

SPENCER GORDON BENNET (1893-)

1929: Hawk of the Hills
1930: Rogue of the Rio Grande
1933: Justice Takes a Holiday; The Midnight Warning
1934: Badge of Honor; Fighting Rookie; Night Alarm; The Oil Raider
1935: Calling All Cars; Get That Man; Heir to Trouble; Lawless Riders; Rescue Squad; Western Courage
1936: Avenging Waters; Heroes of the Range; Ranger Courage; The Cattle Thief; The Fugitive Sheriff; The Unknown Ranger; Rio Grande Ranger
1937: The Law of the Ranger; The Rangers Step In; Reckless Ranger
1939: Oklahoma Terror; Across the Plains; Riders of the Frontier
1940: Cowboy from Sundown; Westbound Stage
1941: Arizona Bound; Gunman from Bodie; Ridin' the Cherokee Trail
1942: They Raid by Night
1943: Calling Wild Bill Elliott; Canyon City
1944: Beneath Western Skies; California Joe; Code of the Prairie; Mojave Firebrand; Tucson Raiders
1945: Lone Texas Ranger
1952: Brave Warrior; Voodoo Tiger
1953: Killer Ape; Savage Mutiny
1955: Devil Goddess
1959: Submarine Seahawk
1960: The Atomic Submarine
1965: Requiem for a Gunfighter; The Bounty Killer

(Serials)

1932: The Last Frontier
1937: The Mysterious Pilot

1942: The Secret Code; The Valley of Vanishing Men
1943: Secret Service in Darkest Africa; The Masked Marvel
1944: Haunted Harbor; The Tiger Woman; Zorro's Black Whip (all w/ W. Grissell)
1945: Manhunt on Mystery Island; Federal Operator 99 (both w/ W. Grissell, Y. Canutt); The Purple Monster Strikes (w/ F. Brannon)
1946: The Phantom Rider; Daughter of Don Q; King of the Forest Rangers (all w/ F. Brannon)
1947: Son of Zorro; The Black Widow (both w/ F. Brannon); Brick Bradford
1948: Superman; Congo Bill (both w/ T. Carr)
1949: Adventures of Sir Galahad; Bruce Gentry--Daredevil of the Skies (w/ T. Carr); Batman and Robin
1950: Cody of the Pony Express; Atom Man vs. Superman; Pirates of the High Seas (w/ T. Carr)
1951: Roar of the Iron Horse (w/ T. Carr); Mysterious Island; Captain Video (w/ W. Grissell)
1952: Blackhawk; Son of Geronimo; King of the Congo (w/ W. Grissell)
1953: The Lost Planet
1954: Gunfighters of the Northwest; Riding with Buffalo Bill
1955: Adventures of Captain Africa
1956: Perils of the Wilderness; Blazing the Overland Trail
1957: Congo Bill, King of the Jungle (w/ T. Carr)

CHARLES BENNETT (1899-   )

1953: No Escape

COMPTON BENNETT (1900-1974) Brit.

(American films only)
1948: My Own True Love
1949: That Forsyte Woman
1950: King Solomon's Mines (w/ A. Marton)

HUGH BENNETT

1941: Henry Aldrich for President
1942: Henry Aldrich, Editor; Henry and Dizzy
1943: Henry Aldrich Gets Glamour; Henry Aldrich Haunts a House; Henry Aldrich Swings It
1944: Henry Aldrich, Boy Scout; Henry Aldrich Plays Cupid; Henry Aldrich's Little Secret; National Barn Dance

RICHARD BENNETT

1978: Harper Valley, P.T.A.

ANDREW V. BENNISON

1932: This Sporting Age (w/ A. F. Erickson)

LEON BENSON

1964: Flipper's New Adventure

ROBERT BENTON (1933-)

1972: Bad Company
1977: The Late Show
1979: Kramer vs. Kramer

LEONARDO BERCOVICI

1970: Story of a Woman

LUDWIG BERGER (1892-1969)

1930: The Vagabond King; Playboy of Paris
1931: Le Petit Cafe (Fr. version of Playboy of Paris)
1940: The Thief of Bagdad (Brit. w/ M. Powell, Tim Whelan)

WILLIAM BERKE (1903-1958)

1935: The Pecos Kid; Toll of the Desert
1936: Desert Justice; Gun Grit
1942: Badmen of the Hills; Down Rio Grande Way; Lawless Plainsmen; Overland to Deadwood; Riders of the Northland; The Lone Prairie; A Tornado in the Saddle
1943: The Fighting Buckaroo; Hail to the Rangers; Law of the Northwest; Pardon My Gun; Riding Through Nevada; Robin Hood of the Range; Tornado; Frontier Fury; Minesweeper; Silver City Raiders; Riders of the Northwest Mounted; Saddles & Sagebrush
1944: Double Exposure; The Falcon in Mexico; Riding West; The Girl in the Case; The Navy Way; Sailor's Holiday; That's My Baby; The Vigilantes Ride; Wyoming Hurricane; The Last Horseman; Dark Mountain; Dangerous Passage
1945: Betrayal from the East; Dick Tracy, Detective; Why Girls Leave Home; High Powered
1946: The Falcon's Adventure; Sunset Pass; Ding Dong Williams
1947: Shoot to Kill; Code of the West; Renegade Girl; Rolling Home
1948: Caged Fury; Jungle Jim; Highway 13; Racing Luck; Speed to Spare; Waterfront at Midnight
1949: The Lost Tribe; Zamba; Deputy Marshal; Arson, Inc.; Sky Liner; Treasure of Monte Cristo

1950: Captive Girl; Gunfire; Pygmy Island; Mark of the Gorilla; On the Isle of Samoa; Operation Haylift; I Shot Billy the Kid; Border Rangers; Bandit Queen; Everybody's Dancing; Train to Tombstone
1951: FBI Girl; Fury of the Congo; Savage Drums; Smuggler's Gold; Danger Zone; Pier 23; Roaring City
1952: The Jungle
1953: The Marshal's Daughter; Valley of the Headhunters
1957: Four Boys and a Gun; Street of Sinners
1958: Cop Hater; Island of Women; The Muggers; The Lost Missile

BUSBY BERKELEY (1895-1976)

1933: She Had to Say Yes (w/ G. Amy)
1935: Bright Lights; Gold Diggers of 1935; I Live for Love
1936: Stage Struck
1937: Hollywood Hotel; The Go Getter
1938: Garden of the Moon; Men Are Such Fools; Comet over Broadway
1939: Babes in Arms; They Made Me a Criminal; Fast and Furious
1940: Forty Little Mothers; Strike Up the Band
1941: Babes on Broadway; Blonde Inspiration
1942: For Me and My Gal; Born to Sing (w/ E. Ludwig)
1943: The Gang's All Here
1946: Cinderella Jones
1949: Take Me Out to the Ball Game

DAVID BERLATSKY

1977: The Farmer

ABBY BERLIN

1945: Leave It to Blondie; Life with Blondie
1946: Blondie Knows Best; Blondie's Lucky Day
1947: Blondie in the Dough; Blondie's Anniversary; Blondie's Big Moment; Blondie's Holiday
1948: Blondie's Reward
1949: Mary Ryan, Detective
1950: Double Deal

MONTY BERMAN

1961: The Secret of Monte Cristo

SHELLEY BERMAN

1975: Keep Off! Keep Off!

EDWARD BERNDS (1911-   )

1948: Blondie's Secret
1949: Blondie's Big Deal; Blondie Hits the Jackpot; Feudin' Rhythm
1950: Beware of Blondie; Blondie's Hero
1951: Corky of Gasoline Alley; Gasoline Alley; Gold Raiders
1952: The Harem Girl
1953: Clipped Wings; Hot News; Loose in London; Private Eyes
1954: Jungle Gents; The Bowery Boys Meet the Monsters
1955: Bowery to Bagdad; Spy Chasers
1956: Calling Homicide; Dig That Uranium; Navy Wife; World Without End
1957: Reform School Girl; The Storm Rider
1958: Escape from Red Rock; Joy Ride; Quantrill's Raiders; Queen of Outer Space; Space Master
1959: Alaska Passage; High School Hellcats; Return of the Fly
1961: Valley of the Dragons
1962: The Three Stooges in Orbit; The Three Stooges Meet Hercules

JOSEPH BERNE

1934: Dawn to Dawn
1944: They Live in Fear
1946: Down Missouri Way

JACK BERNHARD

1946: Decoy; Sweetheart of Sigma Chi; Violence
1947: In Self Defense
1948: Appointment with Murder; The Hunted; Unknown Island; Blond Ice; Perilous Waters
1949: Search for Danger; Alaska Patrol
1950: The Second Face

CURTIS BERNHARDT (1899-1981)

1930: Three Loves (Germ.); Thirteen Men and a Girl (Germ.)
1937: The Beloved Vagabond (Brit.)
1938: The Girl in the Taxi
1940: Lady with Red Hair; My Love Came Back
1941: Million Dollar Baby
1942: Juke Girl
1943: Happy Go Lucky
1945: Conflict
1946: My Reputation; Devotion; A Stolen Life
1947: Possessed; High Wall
1949: The Doctor and the Girl
1951: Payment on Demand; Sirocco; The Blue Veil

1952: The Merry Widow
1953: Miss Sadie Thompson
1954: Beau Brummell
1955: Interrupted Melody
1956: Gaby
1962: Damon and Pythias (Ital.)
1963: Stephanie in Rio
1964: Kisses for My President

JOHN BERRY (1917-)

1945: Miss Susie Slagle's
1946: From This Day Forward; Cross My Heart
1948: Casbah
1949: Tension
1951: He Ran All the Way
1952: C'est Arrivé à Paris (Fr.)
1957: Pantaloons
1959: Tamango (Fr.)
1966: Maya (India)
1974: Claudine
1977: Thieves
1978: The Bad News Bears Go to Japan

IRWIN BERWICK

1958: The Monster of Piedras Biancas
1979: Malibu High

CHRIS BEUTE

1938: The Headleys at Home

RICHARD BEYMER (1939-)

1974: The Interview

ABNER BIBERMAN (1909-1977)

1954: The Golden Mistress
1955: The Looters; Running Wild
1956: Behind the High Wall; The Price of Fear

1957: Gun for a Coward; The Night Runner
1958: Flood Tide

HERBERT J. BIBERMAN (1900-1971)

1935: One Way Ticket
1936: Meet Nero Wolfe
1944: The Master Race
1954: Salt of the Earth
1969: Slaves

BRUCE BILSON

1979: The North Avenue Irregulars

STEVE BINDER

1975: Give 'Em Hell, Harry!

JOSH BINNEY

1933: Across the Rio Grande; My Gypsy Sweetheart; Rangers at War; Where Cattle Is King
1947: The Good Shepherd; Hi Di Ho; The Producer's Dilemma
1948: Boardin' House Blues; Killer Diller

CLAUDE BINYON (1905-1978)

1948: Family Honeymoon; The Saxon Charm
1950: Mother Didn't Tell Me; Stella
1951: Aaron Slick from Punkin' Crick
1952: Dreamboat
1953: Here Come the Girls

ROLLIN BINZER

1974: Ladies and Gentlemen, the Rolling Stones (doc.)

RUSSELL J. BIRDWELL

1929: Masquerade
1933: Flying Devils
1956: The Come-On
1957: The Girl in the Kremlin

SAM BISCHOFF (1890-1974)

1932: The Last Mile

STIG BJORKMAN

1972: Georgia, Georgia

NOEL BLACK (1937-)

1968: Pretty Poison
1970: Cover Me Babe
1971: Jennifer on My Mind
1979: A Man, a Woman and a Bank (Can.)

RICHARD BLACKBURN

1975: The Legendary Curse of Lemora

GEORGE BLAIR (1906-1970)

1944: Silent Partner; Secrets of Scotland Yard; End of the Road
1945: Scotland Yard Investigator; A Sporting Chance
1946: G.I. War Bride; Gay Blades; Affairs of Geraldine; That's My Girl
1947: The Ghost Goes Wild; Exposed; The Trespasser
1948: Daredevils of the Clouds; Homicide for Three; King of the Gamblers; Lightnin' in the Forest; Madonna of the Desert; Show Time
1949: Alias the Champ; Daughter of the Jungle; Duke of Chicago; Flaming Fury; Post Office Investigator; Rose of the Yukon; Streets of San Francisco
1950: Under Mexicali Stars; Unmasked; Federal Agents at Large; Destination Big House; Woman from Headquarters; Lonely Hearts Bandits; The Missourians
1951: Thunder in God's Country; Insurance Investigator; Silver City Bonanza; Secrets of Monte Carlo
1952: Desert Pursuit; Woman in the Dark
1953: Perils of the Jungle
1955: The Twinkle in God's Eye
1956: Fighting Trouble; Jaguar
1957: Sabu and the Magic Ring; Spook Chasers
1960: The Hypnotic Eye

MILTON BLAIR

1967: Surfari (doc.)
1970: Blue Surfari (doc.)

FOLMAR BLANGSTED

1937: The Old Wyoming Trail; Westbound Mail

EDWARD A. BLATT (1905-)

1944: Between Two Worlds
1945: Escape in the Desert
1948: Smart Woman

GEORGE BLOOMFIELD

1970: Jenny
1972: To Kill a Clown

RALPH C. BLUEMKE

1968: Robby

MORT BLUMENSTOCK

1931: Morals for Women

JASPER BLYSTONE

1941: The Reluctant Dragon (anim.)

JOHN G. BLYSTONE (1892-1938)

1929: Thru Different Eyes; The Sky Hawks; Captain Lash
1930: So This Is London; Tol'able David; The Big Party
1931: Mr. Lemon of Orange; Young Sinners; Men on Call
1932: The Painted Woman; Too Busy to Work; Charlie Chan's Chance; She Wanted a Millionaire; Amateur Daddy
1933: Hot Pepper; Shanghai Madness; My Lips Betray
1934: Change of Heart; Coming-Out Party; Hell in the Heavens
1935: The County Chairman; Bad Boy
1936: Great Guy; Gentle Julia; Little Miss Nobody; The Magnificent Brute
1937: $23\frac{1}{2}$ Hours Leave; Woman Chases Man; Music for Madame
1938: Block-Heads; Swiss Miss

AL BOASBERG

1934: Myrt and Marge

FRANK Q. BOBBS

1975: Disciples of Death

BUDD BOETTICHER (1916-)

1944: One Mysterious Night; The Missing Juror
1945: Youth on Trial; A Guy, a Gal and a Pal; Escape in the Fog
1946: The Fleet That Came to Stay
1948: Assigned to Danger; Behind Locked Doors
1949: Black Midnight
1950: Wolf Hunters; Killer Shark
1951: Sword of D'Artagnan; The Bullfighter and the Lady; The Cimarron Kid
1952: Red Ball Express; Bronco Buster; Horizons West
1953: City Beneath the Sea; Seminole; East of Sumatra; Wings of the Hawk; The Man from the Alamo
1955: The Magnificent Matador
1956: The Killer Is Loose; Seven Men from Now
1957: Decision at Sundown; The Tall T
1958: Buchanan Rides Alone
1959: Ride Londsome; Westbound
1960: The Rise and Fall of Legs Diamond; Comanche Station
1968: Arruza
1971: A Time for Dying

PAUL BOGART (1919-)

1969: Marlowe
1970: Halls of Anger
1971: Skin Game
1972: Cancel My Reservation
1973: Class of '44
1975: Mr. Ricco; Tell Me Where It Hurts
1977: The Three Sisters

PETER BOGDANOVICH (1939-)

1966: Voyage to the Planet of Prehistoric Women
1968: Targets
1971: The Last Picture Show
1972: What's Up, Doc?
1973: Paper Moon
1974: Daisy Miller
1975: At Long Last Love
1976: Nickelodeon
1979: Saint Jack

RICHARD BOLESLAWSKI (1889-1937)

1930: Last of the Lone Wolf
1931: Woman Pursued; Gay Diplomat
1932: Rasputin and the Empress; Storm at Daybreak; Beauty for Sale
1934: Fugitive Lovers; Men in White; Operator 13; The Painted Veil
1935: Clive of India; Les Miserables; O'Shaughnessy's Boy; Metropolitan
1936: The Three Godfathers; The Garden of Allah; Theodora Goes Wild
1937: The Last of Mrs. Cheyney (completed by George Fitzmaurice)

JACK BOMAY

1974: Solomon King (w/ Sal Watts)

JOHN BOORMAN (1933-)

1965: Having a Wild Weekend (Brit.)
1967: Point Blank
1968: Hell in the Pacific
1969: Rosencrantz and Guildenstern Are Dead
1970: Leo the Last (Brit.)
1972: Deliverance
1974: Zardoz
1977: Exorcist II: The Heretic

CARLOS BORCOSQUE

1934: The Fighting Lady

FRANK BORZAGE (1894-1962)

1929: They Had to See Paris
1930: Song o' My Heart; Liliom
1931: Doctors' Wives; As Young As You Feel; Bad Girl
1932: A Farewell to Arms; After Tomorrow; Young America
1933: Man's Castle; Secrets
1934: Flirtation Walk; No Greater Glory; Little Man, What Now?
1935: Living on Velvet; Shipmates Forever; Stranded
1936: Desire; Hearts Divided
1937: The Big City (ret. Skyscraper Wilderness); Green Light; History Is Made at Night; Mannequin
1938: The Shining Hour; Three Comrades
1939: Disputed Passage
1940: Flight Command; The Mortal Storm; Strange Cargo

1941: Smilin' Through; The Vanishing Virginian
1942: Seven Sweethearts
1943: His Butler's Sister; Stage Door Canteen
1944: Till We Meet Again
1945: The Spanish Main
1946: I've Always Loved You; Magnificent Doll
1947: That's My Man
1948: Moonrise
1958: China Doll
1959: The Big Fisherman

NED BOSNICK

1970: Imago
1972: To Be Free

ROY BOULTING (1913-) Brit.

(American films only)
1956: Run for the Sun
1979: The Number

CHARLES R. BRABIN (1883-1957)

1929: The Bridge of San Luis Rey
1930: The Ship from Shanghai; Call of the Flesh
1931: The Great Meadow; Sporting Blood
1932: The Mask of Fu Manchu; The Beast of the City; New Morals for Old; The Washington Masquerade
1933: The Secret of Madame Blanche; Stage Mother; Day of Reckoning
1934: A Wicked Woman

BERT BRACKEN

1932: The Face on the Barroom Floor

BASIL BRADBURY

1973: A Taste of Hell (w/ Neil Yarema)

ROBERT NORTH BRADBURY (c.1885-)

1931: Dugan of the Bad Lands; Son of the Plains
1932: Hidden Valley; Man from Hell's Edges; Riders of the Desert; Texas Buddies; Law of the West; Son of Oklahoma

1933: Breed of the Border; Gallant Fool; Galloping Romeo; Ranger's Code
1934: Riders of Destiny; The Star Packer; Blue Steel; The Lucky Texan; The Man from Utah; West of the Divide; Happy Landing
1935: Between Men; Courageous Avenger; The Dawn Rider; The Lawless Frontier; Rainbow Valley; Smokey Smith; Texas Terror; Westward Ho; Kid Courage; Rider of the Law; Western Justice
1936: The Lawless Range; Valley of the Lawless; Cavalry; Headin' for the Rio Grande; The Last of the Warrens; Sundown Saunders
1937: Trouble in Texas; The Gun Ranger; The Trusted Outlaw; Hittin' the Trail; Riders of the Dawn; Riders of the Rockies; Stars over Arizona; Where Trails Divide; Sing, Cowboy, Sing; Danger Valley
1942: Forbidden Trails

DAVID BRADLEY (1920-)

1941: Peer Gynt; Treasure Island
1946: Macbeth
1950: Julius Caesar
1952: Talk About a Stranger
1958: Dragstrip Riot
1960: Twelve to the Moon
1964: Madmen of Mandoras

JOHN BRAHM (1893-)

1935: Scrooge
1936: The Last Journey
1937: Counsel for Crime; Broken Blossoms
1938: Penitentiary; Girls' School
1939: Let Us Live; Rio
1940: Escape to Glory (alt. Submarine Zone)
1941: Wild Geese Calling
1942: The Undying Monster
1943: Tonight We Raid Calais; Wintertime
1944: Guest in the House; The Lodger
1945: Hangover Square
1946: The Locket
1947: The Brasher Doubloon; Singapore
1951: The Thief of Venice
1952: Face to Face (w/ B. Windust); The Miracle of Our Lady of Fatima
1953: The Diamond Queen
1954: The Mad Magician
1955: Bengazi; Special Delivery
1967: Hot Rods to Hell

BILL BRAME

1970: Cycle Savages

MARLON BRANDO (1924-   )

1961: One-Eyed Jacks

FRED BRANNON (1901-   )

1949: Bandit King of Texas; Frontier Investigator
1950: Salt Lake Raiders; Code of the Silver Sage; Gunmen of Abilene; Rustlers on Horseback; Vigilante Hideout
1951: Night Riders of Montana; Rough Riders of Durango; Arizona Manhunt
1952: Captive of Billy the Kid; Wild Horse Ambush
1958: Satan's Satellites; Missile Monsters

(Serials)

1945: The Purple Monster Strikes (w/ S. Bennet)
1946: The Crimson Ghost (w/ W. Witney); Daughter of Don Q (w/ S. Bennet); King of the Forest Rangers (w/ S. Bennet); The Phantom Rider (w/ S. Bennet)
1947: The Black Widow (w/ S. Bennet); Jesse James Rides Again (w/ T. Carr); Son of Zorro (w/ S. Bennet)
1948: Adventures of Frank and Jesse James (w/ Y. Canutt); Dangers of the Canadian Mounted (w/ Y. Canutt); G-Men Never Forget (w/ Y. Canutt)
1949: Federal Agents vs. Underworld, Inc.; Ghost of Zorro; King of the Rocketmen
1950: Desperadoes of the West; The Invisible Monster; The James Brothers of Missouri; Radar Patrol vs. Spy Ring
1951: Don Daredevil Rides Again; Flying Disc Man from Mars; Government Agents vs. Phantom Legion; Pirates Harbor
1953: Jungle Drums of Africa

KEEFE BRASSELLE (1923-   )

1958: The Fighting Wildcats (Brit.)

GEORGE BREAKSTON (1920-1973)

1948: Urubu
1950: Jungle Stampede
1954: The Scarlet Spear
1955: Golden Ivory
1956: Escape in the Sun

1957: The Woman and the Hunter
1966: The Boy Cried Murder (Brit.)
1968: Blood River

IRVING S. BRECHER (1914-)

1949: The Life of Riley
1952: Somebody Loves Me
1962: Sail a Crooked Ship

RICHARD L. BREEN (1919-1967)

1957: Stopover Tokyo

MILTON BREN

1952: Three for Bedroom C

HERBERT BRENON (1880-1958) Brit.

(American films only)
1929: The Rescue
1930: The Case of Sergeant Grischa; Lummox
1931: Beau Ideal; Transgression
1932: Girl of the Rio
1933: Wine, Women and Song

HUGH BRESLOW

1951: You Never Can Tell

MARTIN BREST

1977: Hot Tomorrows
1979: Going in Style

HOWARD BRETHERTON (1896-1969)

1929: From Headquarters; Greyhound Limited; The Time, the Place and the Girl; The Redeeming Sin
1930: Second Choice; Isle of Escape
1932: The Match King
1933: Ladies They Talk About (w/ W. Keighley)
1934: Return of the Terror
1935: Dinky; Bar 20 Rides Again; The Eagle's Brood; Hopalong Cassidy

1936: The Leathernecks Have Landed; Three on the Trail; King of the Royal Mounted; Call of the Prairie; Heart of the West; Girl from Mandalay; Wild Brian Kent; Secret Valley
1937: It Happened Out West; County Fair; Western Gold
1938: Wanted by the Police
1939: Boys' Reformatory; Danger Flight; Irish Luck; Navy Secrets; Sky Patrol; Tough Kid; Undercover Agent
1940: The Showdown; Chasing Trouble; Laughing at Danger; Midnight Limited; On the Spot; Up in the Air
1941: In Old Colorado; Outlaws of the Desert; Twilight on the Trail; Sign of the Wolf; You're Out of Luck
1942: Below the Border; Ghost Town Law; Down Texas Way; Riders of the West; West of the Law; Pirates of the Prairie; Riders of the Badlands; West of Tombstone; Dawn on the Great Divide; Rhythm Parade (w/ D. Gould)
1943: Beyond the Last Frontier; Bordertown Gun Fighters; Wagon Trails West; Whispering Footsteps; Carson City Cyclone; Fugitive from Sonora; Riders of the Rio Grande
1944: Hidden Valley Outlaws; Law of the Valley; The San Antonio Kid; Outlaws of Santa Fe; The Girl Who Dared
1945: Gun Smoke; The Navajo Trail; Renegades of the Rio Grande; The Topeka Terror; The Big Showoff
1947: The Trap; Ridin' Down the Trail
1948: The Prince of Thieves; Triggerman; Where the North Begins
1952: Night Raiders

(Serials)

1945: Who's Guilty? (w/ W. Grissell); The Monster and the Ape

MONTE BRICE

1933: Take a Chance (w/ L. Schwab)
1935: Sweet Surrender

JULES BRICKEN

1957: Drango (w/ H. Bartlett)

CLARENCE BRICKER

1938: Held for Ransom

JAMES BRIDGES

1970: The Baby Maker
1973: The Paper Chase
1977: Sept. 30, 1955
1979: The China Syndrome

RICHARD BRILL

1966: That Tennessee Beat

GEORGE BRITTON

1976: Everyday

ALBERT BROOKS

1979: Real Life

JOSEPH BROOKS

1977: You Light Up My Life
1978: If Ever I See You Again

MEL BROOKS (1926-)

1967: The Producers
1970: The Twelve Chairs
1974: Young Frankenstein; Blazing Saddles
1976: Silent Movie
1977: High Anxiety

RICHARD BROOKS (1912-)

1950: Crisis
1951: The Light Touch
1952: Deadline U.S.A.
1953: Battle Circus; Take the High Ground
1954: The Last Time I Saw Paris; Flame and the Flesh
1955: The Blackboard Jungle
1956: The Catered Affair; The Last Hunt
1957: Something of Value
1958: The Brothers Karamazov; Cat on a Hot Tin Roof
1960: Elmer Gantry
1962: Sweet Bird of Youth
1965: Lord Jim
1966: The Professionals
1967: In Cold Blood
1969: The Happy Ending
1972: $(Dollars)

1975: Bite the Bullet
1977: Looking for Mr. Goodbar

THOR BROOKS

1958: Legion of the Doomed
1959: Arson for Hire

OTTO BROWER (-1946)

1930: The Light of Western Stars (w/ E. H. Knopf); Paramount on Parade (w/ others); The Santa Fe Trail (w/ E. H. Knopf); The Border Legion (w/ E. H. Knopf)
1931: Fighting Caravans (w/ D. Burton); Clearing the Range; The Hard Hombre
1932: Gold; The Local Bad Man; Spirit of the West; Pleasure; Lure of the Sea
1933: Headline Shooter; Crossfire; Scarlet River; Fighting for Justice
1934: Speed Wings; Straightaway; I Can't Escape
1935: The Outlaw Deputy
1936: Sins of Man (w/ G. Ratoff); Postal Inspector
1938: Speed to Burn; Road Demon
1939: Stop, Look and Love; Too Busy to Work; Winner Take All
1940: The Gay Caballero; Youth Will Be Served; Girl from Avenue A; Men with Steel Faces (w/ B. Eason); On Their Own
1942: Little Tokyo, USA
1943: Dixie Dugan
1946: Behind Green Lights

(Serials)

1932: The Devil Horse
1934: Mystery Mountain (w/ B. Eason)
1935: The Phantom Empire (w/ B. Eason)

BARRY BROWN

1970: The Way We Live Now

BRUCE BROWN

1966: The Endless Summer (doc.)
1971: On Any Sunday (doc.)

CLARENCE BROWN (1890-)

1929: Trail of '98; Wonder of Women
1930: Anna Christie; Navy Blues; Romance
1931: A Free Soul; Inspiration; Possessed
1932: The Son-Daughter; Emma; Letty Lynton
1933: Looking Forward; Night Flight
1934: Chained; Sadie McKee
1935: Ah, Wilderness; Anna Karenina
1936: The Gorgeous Hussy; Wife vs. Secretary
1937: Conquest
1938: Of Human Hearts
1939: Idiot's Delight; The Rains Came
1940: Edison, the Man
1941: Come Live with Me; They Met in Bombay
1943: The Human Comedy
1944: National Velvet; The White Cliffs of Dover
1946: The Yearling
1947: Song of Love
1949: Intruder in the Dust
1950: To Please a Lady
1951: Angels in the Outfield; It's a Big Country (w/ others)
1952: Plymouth Adventure; When in Rome

HARRY JOE BROWN (1893-1972)

1929: The Wagon Master; Señor Americano
1930: Parade of the West; The Squealer; The Fighting Legion; Mountain Justice; Song of the Caballero; Sons of the Saddle
1931: A Woman of Experience
1932: Madison Square Garden
1933: Sitting Pretty; The Billion Dollar Scandal; I Love That Man
1944: Knickerbocker Holiday

KARL BROWN (1895-)

1930: Prince of Diamonds (w/ A. H. Van Buren)
1932: Flames
1936: In His Steps; The White Legion
1937: Federal Bullets; Michael O'Halloran; Any Man's Wife
1938: The Port of Missing Girls; Numbered Woman; Barefoot Boy; Under the Big Top

LARRY BROWN

1975: An Eye for an Eye

MELVILLE BROWN (-1938)

1929: Geraldine; Jazz Heaven; The Love Doctor; Dance Hall
1930: Lovin' the Ladies; She's My Weakness; Check and Double Check
1931: Behind Office Doors; White Shoulders; Fanny Foley Herself
1934: Redhead; Lost in the Stratosphere
1935: The Nut Farm; Champagne for Breakfast; Forced Landing
1938: He Loved an Actress (Brit.)

PHIL BROWN

1951: The Harlem Globetrotters

REG BROWN

1953: Son of the Renegade

ROWLAND V. BROWN (1901-1963)

1931: Quick Millions
1932: Hell's Highway
1933: Blood Money

WILLIAM O. BROWN

1969: The Watchmaker

RICOU BROWNING

1975: Salty

TOD BROWNING (1882-1962)

1929: The Thirteenth Chair
1930: Outside the Law
1931: Dracula; Iron Man
1932: Freaks
1933: Fast Workers
1935: Mark of the Vampire
1936: The Devil Doll
1939: Miracles for Sale

S. F. BROWNRIGG

1973: Don't Look in the Basement
1976: Poor White Trash Part 2

JERRY BRUCK, JR.

1973: I. F. Stone's Weekly (doc.)

CLYDE BRUCKMAN (1894-1955)

1930: Feet First
1931: Everything's Rosie
1932: Movie Crazy
1935: Spring Tonic; The Man on the Flying Trapeze

BAIRD BRYANT

1971: Celebration of Big Sur

GERARD BRYANT

1957: Rock Around the World

LARRY BUCHANAN

1963: Free, White and 21
1965: The Eye Creatures
1970: A Bullet for Pretty Boy
1976: Goodbye, Norma Jean

FRANK BUCK (1888-1950)

1935: Fang and Claw (doc.)

BETHEL BUCKALEW

1978: My Boys Are Good Boys

TOM BUCKINGHAM

1932: Cock of the Air

HAROLD S. BUCQUET (1891-1946)

1938: Young Dr. Kildare
1939: Calling Dr. Kildare; On Borrowed Time; The Secret of Dr. Kildare
1940: Dr. Kildare's Strange Case; We Who Are Young; Dr. Kildare Goes Home; Dr. Kildare's Crisis
1941: The Penalty; The People vs. Dr. Kildare; Dr. Kildare's Wedding Day; Kathleen
1942: Calling Dr. Gillespie; The War Against Mrs. Hadley
1943: The Adventures of Tartu (Brit.)
1944: Dragon Seed (w/ Jack Conway)
1945: Without Love

JED BUELL

1940: Mr. Washington Goes to Town

LEO BULGAKOV

1934: White Lies
1935: I'll Love You Always; After the Dance

LUIS BUÑUEL (1900-   ) Span.

(American films only)
1954: Adventures of Robinson Crusoe
1960: The Young One

EDWIN BURKE

1934: Now I'll Tell

PAUL BURNFORD

1945: The Adventures of Rusty

WALTER BURNS

1970: Barbara

DAVID BURTON (1890-   )

1930: The Bishop Murder Case (w/ N. Grinde); Strictly Unconventional
1931: Fighting Caravans (w/ O. Brower); Confessions of a Co-ed (w/ D. Murphy)
1932: Dancers in the Dark
1933: Brief Moment
1934: Lady by Choice; Let's Fall in Love; Sister Under the Skin
1935: Princess O'Hara; The Melody Lingers On
1936: Make Way for a Lady
1940: The Man Who Wouldn't Talk; Manhattan Heartbeat
1941: Private Nurse

JOHN BUSHELMAN

1961: The Silent Call; Sniper's Ridge
1962: The Broken Land
1976: Cat Murkil and the Silks

WILLIAM H. BUSHNELL, JR.

1975: The Four Deuces; Prisoners

DAVID BUTLER (1894-   )

1929: Fox Movietone Follies of 1929; Masked Emotions (w/ Kenneth Hawks); Chasing Through Europe (w/ Alfred L. Werker); Salute (w/ John Ford); Sunny Side Up
1930: High Society Blues; Just Imagine
1931: A Connecticut Yankee; Delicious; Business and Pleasure
1932: Down to Earth; Handle with Care
1933: Hold Me Tight; My Weakness
1934: Bottoms Up; Handy Andy; Have a Heart; Bright Eyes
1935: The Little Colonel; Doubting Thomas; The Littlest Rebel
1936: Captain January; White Fang; Pigskin Parade
1937: Ali Baba Goes to Town; You're a Sweetheart
1938: Kentucky Moonshine; Straight, Place and Show; Kentucky
1939: East Side of Heaven; That's Right--You're Wrong
1940: If I Had My Way; You'll Find Out
1941: Caught in the Draft; Playmates
1942: Road to Morocco
1943: They Got Me Covered; Thank Your Lucky Stars
1944: Shine on Harvest Moon; The Princess and the Pirate
1945: San Antonio
1946: Two Guys from Milwaukee; The Time, the Place, and the Girl
1947: My Wild Irish Rose
1948: Two Guys from Texas
1949: John Loves Mary; Look for the Silver Lining; It's a Great Feeling; The Story of Seabiscuit
1950: The Daughter of Rosie O'Grady; Tea for Two
1951: The Lullaby of Broadway; Painting the Clouds with Sunshine
1952: Where's Charley?; April in Paris
1953: By the Light of the Silvery Moon; Calamity Jane
1954: The Command; King Richard and the Crusaders
1955: Jump into Hell
1956: Glory; The Girl He Left Behind
1961: The Right Approach
1967: C'mon, Let's Live a Little

GEORGE BUTLER

1977: Pumping Iron (doc. w/ Robert Fiore)

ROBERT BUTLER

1969: The Computer Wore Tennis Shoes
1971: The Barefoot Executive; Scandalous John
1972: Now You See Him, Now You Don't

1974: The Ultimate Thrill
1978: Hot Lead and Cold Feet

EDWARD BUZZELL (1897-)

1932: The Big Timer; Hollywood Speaks; Virtue
1933: Child of Manhattan; Ann Carver's Profession; Love, Honor and Oh, Baby!
1934: Cross Country Cruise; The Human Side
1935: The Girl Friend
1936: Transient Lady; Three Married Men; The Luckiest Girl in the World
1937: As Good As Married
1938: Paradise for Three; Fast Company
1939: Honolulu; At the Circus
1940: Go West
1941: The Get-Away; Married Bachelor
1942: Ship Ahoy; The Omaha Trail
1943: The Youngest Profession; Best Foot Forward
1945: Keep Your Powder Dry
1946: Easy to Wed; Three Wise Fools
1947: Song of the Thin Man
1949: Neptune's Daughter
1950: A Woman of Distinction; Emergency Wedding
1953: Confidentially Connie
1955: Ain't Misbehavin'
1961: Mary Had a Little (Brit.)

JOHN BYRUM (1947-)

1976: Inserts

WILLIAM CHRISTY CABANNE (1888-1950)

1930: Conspiracy; The Dawn Trail
1931: The Sky Raiders; Convicted; Graft
1932: Hotel Continental; Midnight Patrol; Hearts of Humanity; The Western Limited; Red-Haired Alibi; The Unwritten Law
1933: Daring Daughters; The World Gone Mad; Midshipman Jack
1934: Money Means Nothing; Jane Eyre; A Girl of the Limberlost; When Strangers Meet
1935: Behind the Green Lights; Rendezvous at Midnight; One Frightened Night; The Keeper of the Bees; Storm over the Andes; Another Face
1936: The Last Outlaw; We Who Are About to Die
1937: Criminal Lawyer; Don't Tell the Wife; The Outcasts of Poker Flat; You Can't Beat Love; Annapolis Salute; The Westland Case

1938: Everybody's Doing It; Night Spot; This Marriage Business
1939: Smashing the Spy Ring; Mutiny on the Blackhawk; Tropic Fury; Legion of Lost Flyers
1940: Man from Montreal; Danger on Wheels; Alias the Deacon; Hot Steel; Black Diamonds; The Mummy's Hand; The Devil's Pipeline
1941: Scattergood Baines; Scattergood Pulls the Strings; Scattergood Meets Broadway
1942: Scattergood Rides High; Drums of the Congo; Timber; Top Sergeant; Scattergood Survives a Murder
1943: Keep 'Em Slugging
1944: Dixie Jamboree
1945: The Man Who Walked Alone
1946: Sensation Hunters
1947: Scared to Death; Robin Hood of Monterey; King of the Bandits
1948: Back Trail; Silver Trails

JAMES CAGNEY (1899-)

1957: Short Cut to Hell

EDWARD L. CAHN (1899-1963)

1931: The Homicide Squad (w/ George Melford)
1932: Law and Order; Radio Patrol; Afraid to Talk
1933: Laughter in Hell; Emergency Call
1935: Confidential
1937: Bad Guy
1941: Redhead
1944: Main Street After Dark
1945: Dangerous Partners
1947: Born to Speed; Gas House Kids in Hollywood
1948: The Checkered Coat; Bungalow 13
1949: I Cheated the Law; Prejudice
1950: The Great Plane Robbery; Destination Murder; Experiment Alcatraz
1951: Two Dollar Bettor
1955: The Creature with the Atom Brain; Betrayed Women
1956: Girls in Prison; The She-Creature; Flesh and the Spur; Runaway Daughters; Shake, Rattle, and Rock
1957: Voodoo Woman; Zombies of Mora-Tau; Dragstrip Girl; Invasion of the Saucer Men; Motorcycle Gang
1958: Jet Attack, Suicide Battalion; It! The Terror from Beyond Space; The Curse of the Faceless Man; Hong Kong Confidential; Guns, Girls, and Gangsters
1959: Riot in Juvenile Prison; Invisible Invaders; The Four Skulls of Jonathan Drake; Pier 5, Havana; Inside the Mafia; Vice Raid

1960: Gunfighters of Abilene; A Dog's Best Friend; Oklahoma Territory; Three Came to Kill; Twelve Hours to Kill; Noose for a Gunman; The Music Box Kid; Cage of Evil; The Walking Target
1961: The Police Dog Story; Frontier Uprising; Operation Bottleneck; Five Guns to Tombstone; The Gambler Wore a Gun; Gun Fight; When the Clock Strikes; You Have to Run Fast; Secret of Deep Harbor; Boy Who Caught a Crook
1962: Incident in an Alley; Gun Street; The Clown and the Kid; Beauty and the Beast

PHIL CAHN

1935: I've Been Around

JERRY CALLAHAN

1935: Gunners and Guns

DONALD CAMMELL

1977: Demon Seed

JOE CAMP (1940-)

1974: Benji
1976: Hawmps!
1977: For Love of Benji
1979: The Double McGuffin

STERLING CAMPBELL

1947: Bush Pilot

MICHAEL CAMPUS

1972: Z. P. G. (Brit.)
1973: The Mack
1974: The Education of Sonny Carson
1976: Survival

RAYMOND CANNON

1929: Why Leave Home?; Joy Street; Red Wine
1930: Ladies Must Play
1931: Night Life in Reno; Swanee River
1932: Hotel Variety
1934: Two Brothers; The Treasure of Wong Low
1937: The Outer Gate; Swing It, Sailor; Behind Prison Bars

YAKIMA CANUTT (1895-)

1945: Sheriff of Cimarron
1948: Carson City Raiders; Oklahoma Badlands; Son of Adventure
1954: The Lawless Rider
1957: Zarak

(Serials)

1945: Manhunt on Mystery Island; Federal Operator 99 (both w/ W. Grissell and S. Bennet)
1948: Adventures of Frank and Jesse James; Dangers of the Canadian Mounted; G-Men Never Forget (all w/ F. Brannon)

LEON CAPETANOS

1974: Summer Run

FRANK CAPRA (1897-)

1929: The Donovan Affair; Flight; The Younger Generation
1930: Ladies of Leisure; Rain or Shine
1931: The Miracle Woman; Platinum Blonde; Dirigible
1932: American Madness; Forbidden
1933: The Bitter Tea of General Yen; Lady for a Day
1934: It Happened One Night; Broadway Bill
1936: Mr. Deeds Goes to Town
1937: Lost Horizon
1938: You Can't Take It with You
1939: Mr. Smith Goes to Washington
1941: Meet John Doe
1944: Arsenic and Old Lace
1946: It's a Wonderful Life
1948: State of the Union
1950: Riding High

1951: Here Comes the Groom
1959: A Hole in the Head
1961: Pocketful of Miracles

LUMAR CARD

1973: The Clones (w/ Paul Hunt)

JACK CARDIFF (1914-   ) Brit.

(American films only)
1962: My Geisha
1963: The Long Ships
1968: Dark of the Sun

JOHN CARDOS

1972: Soul Soldier
1977: Kingdom of the Spiders
1979: The Dark

ANTHONY CARDOZA

1979: Smokey and the Hotwire Gang

EDWIN CAREWE (1883-1940)

1929: Evangeline
1930: The Spoilers
1931: Resurrection
1934: Are We Civilized?

LEWIS JOHN CARLINO

1979: The Great Santini (alt. The Ace)

RICHARD CARLSON (1912-1977)

1954: Four Guns to the Border; Riders to the Stars
1958: Appointment with a Shadow; The Saga of Hemp Brown
1966: Kid Rodelo

HORACE B. CARPENTER

1929: West of the Rockies

JOHN CARPENTER

1974: Dark Star
1976: Assault on Precinct 13
1979: Halloween

BERNARD CARR

1947: Curley; The Fabulous Joe (w/ Harve Foster); The Hal Roach Comedy Carnival
1948: Who Killed "Doc" Robbin?

THOMAS CARR (1907-)

1944: The Cherokee Flash
1945: Santa Fe Saddlemates; Oregon Trail; Bandits of the Badlands; Rough Riders of Cheyenne
1946: Days of Buffalo Bill; The Undercover Woman; Alias Billy the Kid; The El Paso Kid; Red River Renegades; Rio Grande Raiders
1947: Song of the Wasteland; Code of the Saddle
1950: Hostile Country; Marshal of Heldorado; Crooked River; Colorado Ranger; West of the Brazos; Fast of the Draw; Outlaws of Texas
1952: Man from the Black Hills; Wyoming Roundup; The Maverick
1953: The Star of Texas; Rebel City; Topeka; Captain Scarlett; The Fighting Lawman
1954: Bitter Creek; The Forty-Niners; The Desperado
1955: Bobby Ware Is Missing
1956: Three for Jamie Dawn
1957: Dino; The Tall Stranger
1958: Gunsmoke in Tucson
1959: Cast a Long Shadow
1967: Sullivan's Empire (w/ Harvey Hart)

(Serials)

1947: Jesse James Rides Again (w/ F. Brannon)
1948: Superman; Congo Bill (both w/ S. Bennet)
1949: Bruce Genty--Daredevil of the Skies (w/ S. Bennet)
1950: Pirates of the High Seas (w/ S. Bennet)
1951: Roar of the Iron Horse (w/ S. Bennet)
1957: Congo Bill, King of the Jungle (w/ S. Bennet)

DAVID CARRADINE (1936-)

1975: You and Me

ANTHONY CARRAS

1963: Operation Bikini

BARTLETT CARRE

1935: Gunsmoke on the Guadalupe

RICK CARRIER

1962: Strangers in the City

MILTON CARRUTH

1936: Love Letters of a Star (w/ L. R. Foster)
1937: Breezing Home; The Man in Blue; Reported Missing; Some Blondes Are Dangerous; The Lady Fights Back; She's Dangerous (w/ L. R. Foster)

H. P. CARVER

1930: The Silent Enemy

STEVE CARVER

1974: Big Bad Mama
1975: Capone
1976: Drum
1979: Fast Charlie, the Moonbeam Rider

JOHN CASSAVETES (1929-   )

1961: Shadows
1962: Too Late Blues
1963: A Child Is Waiting
1968: Faces
1970: Husbands
1971: Minnie and Moskowitz
1974: A Woman Under the Influence
1976: The Killing of a Chinese Bookie
1979: Opening Night

WILLIAM CASTLE (1914-   )

1943: The Chance of a Lifetime; Klondike Kate
1944: The Whistler; When Strangers Marry; She's a Soldier, Too;

The Mark of the Whistler
1945: Voice of the Whistler; Crime Doctor's Warning
1946: Just Before Dawn; The Mysterious Intruder; The Return of Rusty; The Crime Doctor's Man Hunt
1947: The Crime Doctor's Gamble
1948: Texas, Brooklyn, and Heaven; The Gentleman from Nowhere
1949: Johnny Stool Pigeon; Undertow
1950: It's a Small World
1951: The Fat Man; Hollywood Story; Cave of Outlaws
1953: Serpent of the Nile; Fort Ti; Conquest of Cochise; Slaves of Babylon
1954: Charge of the Lancers; Drums of Tahiti; Jesse James vs. the Daltons; Battle of Rogue River; The Iron Glove; The Saracen Blade; The Law vs. Billy the Kid; Masterson of Kansas
1955: The Americano; New Orleans Uncensored; The Gun That Won the West; Duel on the Mississippi
1956: The Houston Story; Uranium Boom
1958: Macabre; The House on Haunted Hill
1959: The Tingler
1960: 13 Ghosts
1961: Homicidal; Mr. Sardonicus
1962: Zotz!
1963: 13 Frightened Girls; The Old Dark House
1964: Strait-Jacket; The Nightwalker
1965: I Saw What You Did
1966: Let's Kill Uncle
1967: The Busy Body; The Spirit Is Willing
1968: Project X
1974: Shanks

WILLIAM ALLEN CASTLEMAN

1973: Bummer
1975: Johnny Firecloud

GILBERT CATES (1934-)

1967: Rings Around the World (doc.)
1970: I Never Sang for My Father
1973: Summer Wishes, Winter Dreams
1976: One Summer Love (orig. Dragonfly)
1979: The Promise

JOSEPH CATES

1960: Girl of the Night

WILLIAM CAYTON

1965: Knockout

1970: AKA Cassius Clay
1971: Jack Johnson (doc.)

RALPH CEDAR

1932: A Fool's Advice
1934: She Had to Choose
1938: Meet the Mayor
1940: West of Abilene

DON CHAFFEY (1917-) Brit.

(American films only)
1964: The Three Lives of Thomasina
1971: Creatures the World Forgot
1973: Charley One-Eye
1976: Ride a Wild Pony
1977: Pete's Dragon
1978: The Magic of Lassie
1979: C.H.O.M.P.S.

NORMAN C. CHAITIN

1962: The Small Hours

WIN CHAMBERLAIN

1970: Brand X

GOWER CHAMPION (1921-1980)

1963: My Six Loves
1974: The Bank Shot

JOHN CHAMPION (1923-)

1976: Mustang Country

HARRY CHAPIN

1968: The Legendary Champions (doc.)

CHARLES CHAPLIN (1889-1977)

1931: City Lights
1935: Modern Times

1940: The Great Dictator
1947: Monsieur Verdoux
1952: Limelight
1957: A King in New York
1967: A Countess from Hong Kong

ERIK CHARRELL (1895-1974)

1934: Caravan

ROBERT C. CHINN

1976: Panama Red

MARVIN J. CHOMSKY

1971: Evel Knievel
1975: Live a Little, Steal a Lot; Mackintosh and T. J.
1979: Good Luck, Miss Wyckoff

LEON CHOOLUCK (1920-   )

1960: Three Blondes in His Life

CHRIS CHRISTENBERRY

1973: Little Cigars

BENJAMIN CHRISTENSEN (1879-1959) Danish

(American films only)
1929: The Mysterious Island (w/ M. Tournier, L. Hubbard); The House of Horror; Seven Footprints to Satan

JOHN CHRISTIAN

1953: Guerrilla Girl

AL CHRISTIE (1886-1951)

1930: Charley's Aunt
1938: Birth of a Baby
1940: Half a Sinner

HOWARD CHRISTY

1933: Sing, Sinner, Sing

DAVID & BYRON CHUDNOW

1972: The Doberman Gang
1973: The Daring Dobermans
1976: The Amazing Dobermans

MATT CIMBER

1971: Calliope
1975: The Candy Tangerine Man; Lady Cocoa
1976: The Witch Who Came from the Sea

MICHAEL CIMINO (1943-)

1974: Thunderbolt and Lightfoot
1978: The Deer Hunter

RENE CLAIR (1898-1981) Fr.

(American films only)
1941: The Flame of New Orleans
1942: I Married a Witch
1943: Forever and a Day (w/ others)
1944: It Happened Tomorrow
1945: And Then There Were None

BENJAMIN CLARK

1972: Children Shouldn't Play with Dead Things

BOB CLARK

1972: Deathdream
1974: Dead of Night
1976: Breaking Point (Can.)

BRUCE CLARK

1972: Hammer

COLBERT CLARK

(Serials)

1933: Mystery Squadron (w/ David Howard); Fighting with Kit Carson (w/ A. Schaefer); The Three Musketeers (w/ A. Schaefer); The Whispering Shadow (w/ A. Herman); Wolf Dog (w/ H. Fraser)
1934: Burn 'Em Up Barnes (w/ A. Schaefer)

GREYDON CLARK (1943-)

1976: Black Shampoo; The Bad Bunch; Tom

JAMES B. CLARK

1957: Under Fire
1958: Sierra Baron; Villa!
1959: A Dog of Flanders; The Sad Horse
1960: One Foot in Hell
1961: The Big Show; Misty
1963: Flipper; Drums of Africa
1964: Island of the Blue Dolphins
1966: ... And Now Miguel
1969: My Side of the Mountain
1972: The Little Ark
1974: Madhouse

SHIRLEY CLARKE (1925-)

1962: The Connection
1964: Cool World
1967: Portrait of Jason (doc.)

JAMES CLAVELL (1924-)

1959: Five Gates to Hell
1960: Walk Like a Dragon
1962: The Sweet and the Bitter (Can.)
1967: To Sir with Love (Brit.)
1969: Where's Jack? (Brit.)
1971: The Last Valley (Brit.)

WILLIAM F. CLAXTON

1948: Half-Past Midnight
1949: Tucson
1951: All That I Have

1954: Fangs of the Wild
1956: Stagecoach to Fury
1957: God Is My Partner; Young and Dangerous; The Quiet Gun; Rockabye Baby
1960: Desire in the Dust; Young Jesse James
1963: Law of the Lawless
1964: Stage to Thunder Rock
1972: Night of the Lepus

JACK CLAYTON (1921-   ) Brit.

(American films only)
1967: Our Mother's House (Brit./U.S.)
1974: The Great Gatsby

WILLIAM CLEMENS (1905-   )

1936: Man Hunt; The Law in Her Hands; The Case of the Velvet Claws; Down the Stretch; Here Comes Carter!
1937: Once a Doctor; The Case of the Stuttering Bishop; Talent Scout; The Footloose Heiress; Missing Witnesses
1938: Torchy Blane in Panama; Accidents Will Happen; Mr. Chump; Nancy Drew--Detective
1939: Nancy Drew--Reporter; Nancy Drew--Trouble Shooter; Nancy Drew and the Hidden Staircase; The Dead End Kids on Dress Parade
1940: Calling Philo Vance; King of the Lumberjacks; Devil's Island
1941: She Couldn't Say No; Knockout; The Night of January 16th
1942: A Night in New Orleans; Sweater Girl
1943: Lady Bodyguard; The Falcon in Danger; The Falcon and the Co-eds
1944: The Falcon Out West; Crime by Night
1947: The Thirteenth Hour

RENE CLEMENT (1913-   ) Fr.

(American films only)
1958: This Angry Age (Ital./U.S.)
1966: Is Paris Burning? (Fr./U.S.)
1971: The Deadly Trap (Fr./U.S.)
1975: And Hope to Die (Fr./U.S.)

BILL CLIFFORD

1952: Birthright

ELMER CLIFTON (1890-1949)

1935: Captured in Chinatown; Rip Roaring Riley; Skull and Crown; Pals of the Range; Saddle Courage; Fighting Caballero; Rough Riding Rangers; Border Patrol; Cyclone of the Saddle
1936: Wildcat Trooper; Gambling with Souls
1937: Crusade Against Rackets; Assassin of Youth; Death in the Air; Mile a Minute Lover
1938: California Frontier; Crashin' Thru; Law of the Texan; Stranger from Arizona; Wolves of the Sea; Paroled from the Big House; Crime Afloat
1940: Isle of Destiny
1941: Swamp Woman; Hard Guy; City of Missing Girls; I'll Sell My Life
1942: Deep in the Heart of Texas; The Old Chisholm Trail; The Sundown Kid.
1944: Boss of Rawhide; Dead or Alive; Gangsters of the Frontier; Guns of the Law; The Pinto Bandit; Return of the Rangers; Seven Doors to Death; Spook Town; Whispering Skull
1945: Youth Aflame; Marked for Murder
1949: The Judge; Not Wanted

(Serials)

1936: Custer's Last Stand; The Black Coin
1938: The Secret of Treasure Island
1944: Captain America (w/ J. English)

PETER CLIFTON

1970: Popcorn (doc.)
1971: Superstars in Film Concert (doc.)
1976: The Song Remains the Same (doc. w/ Joe Massot)

EDWARD F. CLINE (1892-1961)

1929: Broadway Fever; His Lucky Day; The Forward Pass
1930: In the Next Room; Sweet Mama; Leathernecking; Hook, Line, and Sinker; The Widow from Chicago
1931: Cracked Nuts; The Naughty Flirt; The Girl Habit
1932: Million Dollar Legs
1933: Parole Girl; So This Is Africa
1934: Peck's Bad Boy; The Dude Ranger
1935: When a Man's a Man; The Cowboy Millionaire; It's a Great Life
1936: F-Man
1937: On Again, Off Again; Forty Naughty Girls; High Flyers
1938: Hawaii Calls; Go Chase Yourself; Breaking the Ice; Peck's Bad Boy with the Circus
1940: My Little Chickadee; The Villain Still Pursued Her; The Bank Dick

1941: Meet the Chump; Hello Sucker; Cracked Nuts; Never Give a Sucker an Even Break
1942: Snuffy Smith, the Yard Bird; What's Cookin'?; Private Buckaroo; Give Out, Sisters; Behind the Eight Ball
1943: He's My Guy; Crazy House
1944: Swingtime Johnny; Slightly Terrific; Ghost Catchers; Moonlight and Cactus; Night Club Girl
1945: See My Lawyer; Penthouse Rhythm
1946: Bringing Up Father
1948: Jiggs and Maggie in Society; Jiggs and Maggie in Court (w/ William Beaudine)

ROBERT CLOUSE

1970: Darker Than Amber; Dreams of Glass
1973: Enter the Dragon
1974: Black Belt Jones; Golden Needles
1975: The Ultimate Warrior
1977: The Pack
1978: The Amsterdam Kill
1979: Game of Death

HAROLD CLURMAN (1901-)

1946: Deadline at Dawn

LEWIS COATES

1979: Starcrash

STEVE COCHRAN (1917-1965)

1967: Tell Me in the Sunlight

GEORGE COCHRANE

1931: Mystery of Life

FRED COE (1914-1979)

1965: A Thousand Clowns
1969: Me, Natalie

BENNETT COHEN

1931: Law of the Rio Grande (w/ Forrest Sheldon)
1934: Rainbow Riders

LARRY COHEN

1972: Bone
1973: Black Caesar; Hell Up in Harlem
1974: It's Alive
1976: God Told Me To
1977: Demon; The Private Files of J. Edgar Hoover
1978: It Lives Again

MARTIN B. COHEN

1967: Rebel Rousers

RICHARD COHEN

1975: Hurry Tomorrow (doc.)

THOMAS COLCHART

1963: Battle Beyond the Sun (Russian dubbed)

C. C. COLEMAN, JR.

1934: Voice in the Night
1936: Legion of Terror; Code of the Range; Dodge City Trail
1937: Parole Racket; Criminals of the Air; Paid to Dance; The Shadow; A Fight to the Finish
1938: When G-Men Step In; Highway Patrol; Squadron of Honor; Flight to Fame
1939: Homicide Bureau; Missing Daughters; Outpost of the Mounties; Spoilers of the Range; My Son Is a Criminal

HERBERT COLEMAN

1961: Battle of Bloody Beach; Posse from Hell

RICHARD A. COLLA

1970: Zigzag
1972: Fuzz
1978: Olly, Olly, Oxen Free (alt. The Great Balloon Adventure)
1979: Battlestar Galactica

BILL COLLERAN

1964: Hamlet

JAMES F. COLLIER

1975: The Hiding Place

ARTHUR GREVILLE COLLINS (1897-1980)

1935: Personal Maid's Secret; The Widow of Monte Carlo
1936: Thank You Jeeves; Nobody's Fool
1937: Paradise Isle
1938: Saleslady

GUNTHER COLLINS

1971: Jud

JUDY COLLINS

1974: Antonia: A Portrait of a Woman (doc. w/ Jill Godmilow)

LEWIS D. COLLINS (1889-1954)

1930: Devil's Pit; Young Desire
1931: Law of the Tong
1933: Gun Law; Via Pony Express; Trouble Busters; Skyways; Ship of Wanted Men
1934: Sing Sing Nights; The Man from Hell; Public Stenographer; Brand of Hate; Ticket to a Crime
1935: Make a Million; The Desert Trail; Hoosier Schoolmaster; Spanish Cape Mystery; Manhattan Butterfly; Along Came a Woman
1936: Down to the Sea; Return of Jimmie Valentine; The Leavenworth Case; Doughnuts and Society; Timber Wolves
1937: Fury and the Woman; Trapped by G-Men; Under Suspicion; The Wildcatter; River of Missing Men; Mighty Treve
1938: Making the Headlines; Flight into Nowhere; Reformatory; Crime Takes a Holiday; The House of Mystery; Outside the Law; The Strange Case of Dr. Meade
1939: Hidden Power; Fugitive at Large; Trapped in the Sky; Whispering Enemies
1940: Outside the Three-Mile Limit; Passport to Alcatraz; Prison Camp; The Great Plane Robbery
1941: The Great Swindle; Borrowed Hero
1942: Little Joe, the Wrangler; Danger in the Pacific
1943: Raiders of San Joaquin; Tenting Tonight on the Old Camp Ground
1944: The Old Texas Trail; Trigger Trail; Sweethearts of the USA; Oklahoma Raiders
1946: Danger Woman

1947: Heading for Heaven; Killer Dill
1948: Jungle Goddess
1949: Fighting Redhead; Ride, Ryder, Ride; Thunder Roll
1950: Cherokee Uprising; Law of the Panhandle; Cowboy and the Prizefighter; Hot Rod
1951: Canyon Raiders; Nevada Badmen; Abilene Trail; Lawless Cowboys; Stage from Blue River; Colorado Ambush; The Longhorn; Man from Sonora; Oklahoma Justice; Texas Lawman
1952: Dead Man's Trail; Fargo; Kansas Territory; Texas City; Waco; Wild Stallion; The Gunman; Montana Incident; Texas Marshal
1953: The Marksman; Canyon Ambush; The Homesteaders; Texas Badman; Vigilante Terror
1954: Two Guns and a Badge

(Serials)

1942: Junior G-Men of the Air (w/ Ray Taylor)
1943: Adventures of the Flying Cadets; The Adventures of Smilin' Jack; Don Winslow of the Coast Guard (all w/ R. Taylor)
1944: The Great Alaskan Mystery; Mystery of the River Boat; Raiders of Ghost City (all w/ R. Taylor)
1945: Jungle Queen; The Master Key; The Royal Mounted Rides Again; Secret Agent X-9 (all w/ R. Taylor)
1946: Lost City of the Jungle; The Scarlet Horseman (both w/ R. Taylor); Mysterious Mr. M. (w/ Vernon Keays)

ROBERT COLLINS

1979: Walk Proud

WALTER COLMES

1945: Identity Unknown; The Woman Who Came Back
1946: The French Key
1947: The Burning Cross; Road to the Big House

GERALD COMIER

1974: Terror Circus

DAVID COMMONS

1969: The Angry Breed

RICHARD COMPTON

1972: Welcome Home, Soldier Boys

1974: Macon County Line; Angels Die Hard (orig. released 1970)
1975: Return to Macon County
1979: The Ravagers

MERLE W. CONNELL

1952: Untamed Woman
1956: The Flesh Merchants

MARC CONNELLY (1890-1980)

1936: The Green Pastures (w/ W. Keighley)

RAY CONNOLLY

1976: James Dean, the First American Teenager (doc.)

ROBERT (BOBBY) CONNOLLY

1937: The Devil's Saddle Legion; Expensive Husbands
1938: The Patient in Room 18 (w/ C. Wilbur)

JACK CONRAD

1975: Country Blue

MIKEL CONRAD

1950: The Flying Saucer

WILLIAM CONRAD (1920-)

1964: The Man from Galveston
1965: Two on a Guillotine; My Blood Runs Cold; Brainstorm

JOHN W. CONSIDINE, JR. (1898-)

1932: Disorderly Conduct

RICHARD CONTE (1914-1975)

1969: Operation Cross Eagles

JACK CONWAY (1887-1952)

1928: Alias Jimmy Valentine; Bringing Up Father
1929: Untamed; Our Modern Maidens
1930: The Unholy Three; New Moon; They Learned About Women (w/ S. Wood)
1931: The Easiest Way; Just a Gigolo
1932: Arsene Lupin; Red Headed Woman; But the Flesh Is Weak
1933: Hell Below; The Nuisance; The Solitaire Man
1934: The Gay Bride; The Girl from Missouri; Tarzan and His Mate (w/ C. Gibbons); Viva Villa!
1935: A Tale of Two Cities; One New York Night
1936: Libeled Lady
1937: Saratoga
1938: Too Hot to Handle; A Yank at Oxford
1939: Lady of the Tropics; Let Freedom Ring
1940: Boom Town
1941: Honky Tonk; Love Crazy
1942: Crossroads
1943: Assignment in Brittany
1944: Dragon Seed
1947: High Barbaree; The Hucksters
1948: Julia Misbehaves

JAMES L. CONWAY

1976: In Search of Noah's Ark (doc.); The Lincoln Conspiracy
1978: Beyond and Back
1979: The Fall of the House of Usher

FIELDER COOK (1923-)

1956: Patterns
1966: A Big Hand for the Little Lady
1968: How to Save a Marriage--and Ruin Your Life; Prudence and the Pill (Brit.)
1971: Eagle in a Cage (Brit.)
1973: From the Mixed-Up Files of Mrs. Basil E. Frankweiler

JACKIE COOPER (1921-)

1972: Stand Up and Be Counted

MERIAN C. COOPER (1893-1973)

1929: Four Feathers (w/ E. Schoedsack, L. Mendes)
1933: King Kong (w/ E. Schoedsack)

JODIE COPELAN

1958: Ambush at Cimarron Pass

JACK L. COPELAND

1957: Hell's Five Hours

FRANCIS FORD COPPOLA (1939-   )

1963: Dementia 13
1967: You're a Big Boy Now
1968: Finian's Rainbow
1969: The Rain People
1972: The Godfather
1974: The Conversation; The Godfather Part II
1979: Apocalypse Now

ROGER (WILLIAM) CORMAN (1926-   )

1955: Five Guns West; Apache Woman
1956: The Day the World Ended; Swamp Woman; The Gunslinger; The Oklahoma Woman; It Conquered the World
1957: Naked Paradise; Attack of the Crab Monsters; Not of This Earth; The Undead; Rock All Night; Carnival Rock; Teenage Doll; Sorority Girl; The Viking Women and the Sea Serpent
1958: War of the Satellites; Machine Gun Kelly; Teenage Caveman; The She-Gods of Shark Reef
1959: I, Mobster; The Wasp Woman; A Bucket of Blood
1960: Ski Troop Attack; The House of Usher; The Little Shop of Horrors; The Last Woman on Earth; Creature from the Haunted Sea; Atlas
1961: The Pit and the Pendulum
1962: Premature Burial; I Hate Your Guts (The Intruder); Tales of Terror; Tower of London
1963: The Raven; The Young Racers; The Haunted Palace; The Terror; X--The Man with X-Ray Eyes
1964: The Masque of the Red Death (Brit.); The Secret Invasion (Yugoslavia)
1965: The Tomb of Ligeia (Brit.)
1966: The Wild Angels
1967: The St. Valentine's Day Massacre; The Trip
1970: Bloody Mama; Gas-s-s-s!
1971: Von Richthofen and Brown (Brit./Ire.)

HUBERT CORNFIELD (1929-   )

1955: Sudden Danger

1957: Lure of the Swamp; Plunder Road
1960: The Third Voice
1961: Angel Baby (w/ P. Wendkos)
1962: Pressure Point
1969: The Night of the Following Day

HAROLD CORNSWEET

1975: Return to Campus

LLOYD CORRIGAN (1900-1969)

1930: Follow Thru (w/ L. Schwab); Along Came Youth (w/ N. McLeod)
1931: Daughter of the Dragon; The Beloved Bachelor
1932: No One Man; The Broken Wing; He Learned About Women
1934: By Your Leave
1935: Murder on a Honeymoon
1936: Dancing Pirate
1937: Night Key; Lady Behave

RICARDO CORTEZ (1899-1977)

1938: The Inside Story
1939: The Escape; Chasing Danger
1940: Heaven With a Barbed Wire Fence; City of Chance; Free, Blonde and 21; Girl in 313

DON COSCARELLI

1976: Jim--The World's Greatest; Kenny and Co.
1979: Phantasm

JOHN COTTER

1979: Mountain Family Robinson

JACK COUFFER

1961: Nikki, Wild Dog of the North
1969: Ring of Bright Water
1972: The Darwin Adventure

WILL COWAN

1958: The Big Beat; The Thing That Couldn't Die

WILLIAM J. COWEN (1883-1964)

1932: Kongo
1933: Oliver Twist
1934: Woman Unafraid

JOHN T. COYLE

1938: Call of the Yukon (w/ B. R. Eason)

GARY L. CRABTREE

1974: Gettin' Back

WILLIAM JAMES CRAFT

1929: The Cohens and the Kellys in Atlantic City; Skinner Steps Out
1930: The Cohens and the Kellys in Scotland; Dames Ahoy!; See America Thirst; Czar of Broadway; Embarrassing Moments; Little Accident; One Hysterical Night
1931: Honeymoon Lane; The Runaround

WILLIAM CRAIN

1972: Blacula
1976: Dr. Black Mr. Hyde
1979: The Watts Monster

EDDIE CRANDALL

1969: From Nashville with Music

KENNETH CRANE

1958: When Hell Broke Loose

FRANK CRAVEN (1875-1945)

1934: That's Gratitude

MICHAEL CRICHTON (1942-)

1973: Westworld
1978: Coma
1979: The Great Train Robbery

DONALD CRISP (1880-1974)

1930: The Runaway Bride

JOHN CROMWELL (1888-1979)

1929: Close Harmony; The Dance of Life; The Mighty
1930: Street of Chance; The Texan; Tom Sawyer; For the Defense; Seven Days' Leave (w/ R. Wallace)
1931: Scandal Sheet; Unfaithful; The Vice Squad; Rich Man's Folly
1932: The World and the Flesh
1933: Sweepings; The Silver Cord; Double Harness; Ann Vickers
1934: The Fountain; Of Human Bondage; Spitfire; This Man Is Mine
1935: I Dream Too Much; Jalna; Village Tale
1936: Banjo on My Knee; Little Lord Fauntleroy; To Mary--With Love
1937: The Prisoner of Zenda
1938: Algiers
1939: In Name Only; Made for Each Other
1940: Abe Lincoln in Illinois; Victory
1941: So Ends Our Night
1942: Son of Fury
1944: Since You Went Away
1945: The Enchanted Cottage
1946: Anna and the King of Siam
1947: Dead Reckoning; Night Song
1950: Caged; The Company She Keeps
1951: The Racket
1958: The Goddess
1961: A Matter of Morals
1963: The Scavengers

GEORGE J. CRONE

1930: Reno; What a Man!; Blaze o' Glory (w/ Renaud Hoffman)
1931: Get That Girl; Speed Madness

DAVID CRONENBERG

1976: They Came from Within

ALAN CROSLAND (1894-1936)

1927: The Jazz Singer
1929: On with the Show; General Crack
1930: The Furies; Son of the Flame; Big Boy; Viennese Nights
1931: Captain Thunder; Children of Dreams
1932: The Silver Lining; Week Ends Only
1934: Massacre; Midnight Alibi; The Personality Kid; The Case of the Howling Dog
1935: The White Cockatoo; It Happened in New York; Mr. Dyna-

mite; Lady Tubbs; King Solomon of Broadway; The Great Impersonation

LAWRENCE CROWLEY

1976: In Search of Bigfoot

WILLIAM X. CROWLEY

1949: Trail of the Yukon

OWEN CRUMP

1962: The Couch

JAMES CRUZE (1884-1942)

1929: The Great Gabbo; The Duke Steps Out; A Man's Man
1930: Once a Gentleman; She Got What She Wanted
1931: Salvation Nell
1932: If I Had a Million (w/ others); Washington Merry-Go-Round
1933: I Cover the Waterfront; Mr. Skitch; Sailor Be Good; Racetrack
1934: David Harum; Their Big Moment
1935: Helldorado; Two Fisted
1936: Sutter's Gold
1937: The Wrong Road
1938: Prison Nurse; Gangs of New York; Come On, Leathernecks

GEORGE CUKOR (1899-)

1930: The Royal Family of Broadway (w/ C. Gardner); The Virtuous Sin; Grumpy
1931: Tarnished Lady; Girls About Town
1932: One Hour with You (w/ E. Lubitsch); A Bill of Divorcement; Rockabye; What Price Hollywood
1933: Dinner at Eight; Little Women; Our Betters
1935: David Copperfield; Sylvia Scarlett
1936: Romeo and Juliet
1937: Camille
1938: Holiday
1939: Zaza; The Women
1940: The Philadelphia Story; Susan and God
1941: Two-Faced Woman; A Woman's Face
1942: Her Cardboard Lover; Keeper of the Flame
1944: Gaslight; Winged Victory

1947: Desire Me
1948: A Double Life
1949: Adam's Rib; Edward, My Son (Brit.)
1950: Born Yesterday; A Life of Her Own
1951: The Model and the Marriage Broker
1953: The Actress
1954: It Should Happen to You; A Star Is Born
1956: Bhowani Junction
1957: Les Girls; Wild Is the Wind
1960: Heller in Pink Tights; Let's Make Love; Song Without End (w/ C. Vidor)
1962: The Chapman Report
1964: My Fair Lady
1969: Justine
1972: Travels with My Aunt
1976: The Blue Bird

ROBERT CULP (1930-)

1972: Hickey and Boggs

EUGENE CUMMINGS

1936: The Crime Patrol

IRVING CUMMINGS (1888-1959)

1929: In Old Arizona (w/ Raoul Walsh); Behind the Curtain; Not Quite Decent
1930: Cameo Kirby; On the Level; A Devil with Women
1931: A Holy Terror; The Cisco Kid
1932: Attorney for the Defense; The Night Club Lady; Man Against Woman
1933: Man Hunt; The Woman I Stole; The Mad Game
1934: I Believed in You; Grand Canary; The White Parade
1935: It's a Small World; Curly Top
1936: The Poor Little Rich Girl; Girls' Dormitory; White Hunter
1937: Vogues of 1938; Merry-Go-Round of 1938
1938: Little Miss Broadway; Just Around the Corner
1939: The Story of Alexander Graham Bell; Hollywood Cavalcade; Everything Happens at Night
1940: Lillian Russell; Down Argentine Way
1941: That Night in Rio; Belle Starr; Louisiana Purchase
1942: My Gal Sal; Springtime in the Rockies
1943: Sweet Rosie O'Grady; What a Woman!
1944: The Impatient Years
1945: The Dolly Sisters
1951: Double Dynamite

RICHARD CUNHA

1958: She Demons; Frankenstein's Daughter; Missile to the Moon
1961: Girl in Room 13

SEAN S. CUNNINGHAM

1971: Together
1974: The Case of the Smiling Stiffs
1978: Here Come the Tigers

JACK CURREY

1971: Run the Wild River (doc.)

DAN CURTIS (1932?-)

1970: House of Dark Shadows
1971: Night of Dark Shadows
1976: Burnt Offerings

MICHAEL CURTIZ (1888-1962)

1929: Noah's Ark; The Glad Rag Doll; Madonna of Avenue A; The Gamblers; Hearts in Exile
1930: Mammy; Under a Texas Moon; The Matrimonial Bed
1931: The Mad Genius; Bright Lights; River's End; God's Gift to Women; The Woman from Monte Carlo; Soldier's Plaything
1932: Cabin in the Cotton; Doctor X; The Strange Love of Molly Louvain; Alias the Doctor
1933: Kennel Murder Case; Mystery of the Wax Museum; 20,000 Years in Sing Sing; The Keyhole; Private Detective 62; Female; Goodbye Again
1934: British Agent; Jimmy the Gent; Mandalay; The Key
1935: Black Fury; Captain Blood; Front Page Woman; The Case of the Curious Bride; Little Big Shot
1936: The Charge of the Light Brigade; The Walking Dead
1937: Kid Galahad; Stolen Holiday; Mountain Justice; The Perfect Specimen
1938: The Adventures of Robin Hood; Four Daughters; Four's a Crowd; Gold Is Where You Find It; Angels with Dirty Faces
1939: Daughters Courageous; Dodge City; Four Wives; The Private Lives of Elizabeth and Essex
1940: Santa Fe Trail; The Sea Hawk; Virginia City
1941: Dive Bomber; The Sea Wolf
1942: Captains of the Clouds; Casablanca; Yankee Doodle Dandy
1943: Mission to Moscow; This Is the Army
1944: Janie; Passage to Marseilles

1945: Mildred Pierce; Roughly Speaking
1946: Night and Day
1947: Life with Father; The Unsuspected
1948: Romance on the High Seas
1949: Flamingo Road; The Lady Takes a Sailor; My Dream Is Yours
1950: The Breaking Point; Bright Leaf; Young Man with a Horn
1951: Force of Arms; I'll See You in My Dreams; Jim Thorpe--All American
1952: The Story of Will Rogers
1953: The Jazz Singer; Trouble Along the Way
1954: The Boy From Oklahoma; The Egyptian; White Christmas
1955: We're No Angels
1956: The Best Things in Life Are Free; The Scarlet Hour; The Vagabond King
1957: The Helen Morgan Story
1958: King Creole; The Proud Rebel
1959: The Hangman; The Man in the Net
1960: The Adventures of Huckleberry Finn; A Breath of Scandal
1961: The Comancheros; Francis of Assisi

RENEE DAALDER

1976: Massacre at Central High

MORTON DA COSTA (1914-)

1958: Auntie Mame
1962: The Music Man
1963: Island of Love

HAROLD DANIELS

1948: The Woman from Tangier
1949: Daughter of the West
1951: Roadblock; The Prince of Peace (w/ W. Beaudine)
1953: Port Sinister; Sword of Venus
1956: Bayou (w/ E. I. Fessler)

MARC DANIELS

1957: The Big Fun Carnival

HERBERT DANSKA

1967: Sweet Love, Bitter
1971: Right On!

JOE DANTE

1978: Piranha

HELMUT DANTINE (1918-)

1958: Thundering Jets

RAY DANTON (1931-)

1972: The Deathmaster
1973: Crypt of the Dead
1975: Psychic Killer

PHILIP D'ANTONI

1973: The Seven-Ups

DANNY DARE

1938: The Main Event

JOAN DARLING (1935-)

1977: First Love

HARRY D'ARRAST (1897-1968)

1930: Laughter
1933: Topaze
1935: The Three-Cornered Hat

JULES DASSIN (1911-)

(American films only)
1942: Reunion in France; Nazi Agent; The Affairs of Martha; Once Upon a Thursday
1943: Young Ideas
1944: The Canterville Ghost
1945: A Letter for Evie
1946: Two Smart People
1947: Brute Force
1948: The Naked City
1949: Thieves' Highway
1950: Night and the City
1960: Never on Sunday (Gr.); Where the Hot Wind Blows (Ital.)

1964: Topkapi
1968: Uptight!; Survival

HERSCHEL DAUGHERTY

1958: The Light in the Forest
1963: The Raiders

JOHN DAUMERY

1930: Rough Waters

DOROTHY DAVENPORT (see MRS. WALLACE REID)

DELMER DAVES (1904-   )

1943: Destination Tokyo
1944: Hollywood Canteen; The Very Thought of You
1945: Pride of the Marines
1947: The Red House; Dark Passage
1948: To the Victor
1949: A Kiss in the Dark; Task Force
1950: Broken Arrow
1951: Bird of Paradise
1952: Return of the Texan
1953: Never Let Me Go; Treasure of the Golden Condor
1954: Demetrius and the Gladiators; Drum Beat
1956: Jubal; The Last Wagon
1957: 3:10 to Yuma
1958: The Badlanders; Cowboy; Kings Go Forth
1959: The Hanging Tree; A Summer Place
1961: Parrish; Susan Slade
1962: Rome Adventure
1963: Spencer's Mountain
1964: Youngblood Hawke
1965: The Battle of the Villa Fiorita

CHARLES DAVID

1945: Lady on a Train; River Gang

HAROLD DAVID

1959: Career Girl

CARSON DAVIDSON

1975: The Wrong Damn Film

E. ROY DAVIDSON

1935: The Best Man Wins (w/ E. C. Kenton)

GORDON DAVIDSON

1972: The Trial of the Catonsville Nine

MARTIN DAVIDSON

1974: The Lords of Flatbush (w/ S. F. Verona)
1978: Almost Summer

VALENTINE DAVIES (1905-)

1955: The Benny Goodman Story
1956: The Goddess

ALLAN DAVIS

1953: Rogue's March

CHARLES DAVIS

1962: Get Out of Town
1973: Happy As the Grass Was Green
1978: Hazel's People

EDDIE DAVIS

1967: Panic in the City
1969: Color Me Dead; It Takes All Kinds

OSSIE DAVIS (1917-)

1970: Cotton Comes to Harlem
1972: Black Girl
1973: Gordon's War; Kongi's Harvest
1976: Countdown at Kusini (Nig.)

PETER DAVIS

1974: Hearts and Minds (doc.)

ROBERT W. DAVISON

1979: Cry to the Wind

TITO DAVISON

1969: The Big Cube

NORMAN DAWN (c.1905-)

1936: Tundra
1939: Taku
1949: Arctic Fury
1950: Two Lost Worlds

RALPH DAWSON (1897-)

1929: The Girl in the Glass Cage

ROBERT DAY (1922-) Brit.

(American films only)
1963: Tarzan's Three Challenges
1966: Tarzan and the Valley of Gold
1967: Tarzan and the Great River

LYMAN D. DAYTON

1976: Baker's Hawk
1979: Rivals

EMILE de ANTONIO (1920-)

1964: Point of Order! (doc.)
1967: Rush to Judgment (doc.)
1968: America Is Hard to See (doc.)
1969: In the Year of the Pig (doc.)
1971: Millhouse: A White Comedy (doc.)
1973: Painters Painting (doc.)
1977: Underground (doc.)

WILLIAM DEAR

1976: The Northville Cemetery Massacre

JOHN DE BELLO

1978: Attack of the Killer Tomatoes

FREDERICK DE CORDOVA (1910-)

1945: Too Young to Know
1946: Her Kind of Man
1947: That Way with Women; Love and Learn; Always Together
1948: Wallflower; For the Love of Mary; The Countess of Monte Cristo
1949: Illegal Entry; The Gal Who Took the West
1950: Buccaneer's Girl; Peggy; The Desert Hawk
1951: Bedtime for Bonzo; Katie Did It; Little Egypt; Finders Keepers
1952: Here Come the Nelsons; Bonzo Goes to College; Yankee Buccaneer
1953: Column South
1965: I'll Take Sweden
1966: Frankie and Johnny

LEANDER DE CORDOVA

1930: After the Fog
1931: Trails of the Golden West

MICHAEL A. DE GAETANO

1974: UFO: Target Earth

EDWARD DEIN

1955: Shack Out on 101
1957: Calypso Joe
1958: Seven Guns to Mesa
1959: Curse of the Undead
1960: The Leech Woman

ROBERT DE LACEY

1929: Idaho Red
1930: Pardon My Gun

EUGENIO DE LAGOURO

1951: Stop That Cab

MERVILLE DE LAY

1944: Law of the Saddle

GERRY DE LEON

1959: Terror Is a Man
1964: The Walls of Hell
1972: Women in Cages

JEAN DE LIMUR

1929: The Letter; Jealousy

ROY DEL RUTH (1895-1961)

1929: Conquest; The Desert Song; The Hottentot; Gold Diggers of Broadway; The Aviator
1930: Hold Everything; The Second Floor Mystery; Three Faces East; The Life of the Party
1931: My Past; Divorce Among Friends; The Maltese Falcon; Side Show; Blonde Crazy; Larceny Lane
1932: Taxi!; Beauty and the Boss; Winner Take All; Blessed Event
1933: Employees Entrance; The Mind Reader; The Little Giant; Captured!; Bureau of Missing Persons; Lady Killer
1934: Bulldog Drummond Strikes Back; Upper World; Kid Millions
1935: Follies Bergere; Broadway Melody of 1936; Thanks a Million!
1936: It Had to Happen; Private Number; Born to Dance
1937: On the Avenue; Broadway Melody of 1938
1938: Happy Landing; My Lucky Star
1939: Tail Spin; The Star Maker; Here I Am a Stranger
1940: He Married His Wife
1941: Topper Returns; The Chocolate Soldier
1942: Maisie Gets Her Man
1943: DuBarry Was a Lady
1944: Broadway Rhythm; Barbary Coast Gent
1947: It Happened on Fifth Avenue
1948: The Babe Ruth Story
1949: Red Light; Always Leave Them Laughing
1950: The West Point Story
1951: On Moonlight Bay; Starlift
1952: About Face: Stop, You're Killing Me
1953: Three Sailors and a Girl
1954: Phantom of the Rue Morgue
1959: The Alligator People

DOM DE LUISE

1979: Hot Stuff

CECIL B. DE MILLE (1881-1959)

1929: Dynamite; The Godless Girl
1930: Madam Satan
1931: The Squaw Man
1932: The Sign of the Cross
1933: This Day and Age
1934: Cleopatra; Four Frightened People
1935: The Crusades
1936: The Plainsman
1938: The Buccaneer
1939: Union Pacific
1940: Northwest Mounted Police
1942: Reap the Wild Wind
1944: The Story of Dr. Wassell
1947: Unconquered
1949: Samson and Delilah
1952: The Greatest Show on Earth
1956: The Ten Commandments

WILLIAM C. DE MILLE (1878-1955)

1929: The Doctor's Secret; The Idle Rich
1930: This Mad World; Passion Flower
1932: Two Kinds of Women
1934: His Double Life (w/ A. Hopkins)

NORMAN DEMING

1939: Riders of Black River; Taming of the West

(Serials)

1939: Mandrake the Magician; Overland with Kit Carson (both w/ Sam Nelson)

JONATHAN DEMME (1944-)

1974: Caged Heat
1975: Crazy Mama
1976: Fighting Mad
1977: Handle with Care
1979: Last Embrace

JACQUES DEMY (1931-   ) Fr.

(American films only)
1969: Model Shop

ARMAND DENIS

1934: Wild Cargo (doc.)
1941: Frank Buck's Jungle Cavalcade (doc.)
1953: Below the Sahara (doc.)

CRAIG DENNEY

1975: The Astroloxer

REGINALD DENNY

1933: The Big Bluff

BRIAN DE PALMA (1941-   )

1968: Murder à la Mod; Greetings
1969: The Wedding Party (w/ others)
1970: Hi Mom!
1972: Get to Know Your Rabbit
1973: Sisters
1974: Phantom of the Paradise
1976: Obsession; Carrie
1978: The Fury

ALLESIO DE PAOLA

1969: Chastity

JOHN DEREK

1965: Once Before I Die
1969: A Boy ... a Girl; Childish Things
1972: Confessions of Tom Harris (w/ David Nelson)

LOUIS DE ROCHEMONT (1889-1978)

1940: The Ramparts We Watch (semi-doc.)
1942: We Are the Marines (doc.)
1958: Windjammer (doc.)

ROBERT C. DERTANO

1957: Journey to Freedom

MARCEL DE SANO

1930: Peacock Alley

VITTORIO DE SICA (1902-1974) Ital.

(American films only)
1954: Indiscretion of an American Wife

JAMES DESMOND

1969: Monterey Pop (w/ others)

STEVE DE SOUZA

1973: Arnold's Wrecking Company

DAVID DETIEGE

1965: The Man from Button Willow (anim.)

ANDRE DE TOTH (c. 1900-   ) Hung.

(American films only)
1943: Passport to Suez
1944: None Shall Escape; Dark Waters
1947: Ramrod; The Other Love
1948: The Pitfall
1949: Slattery's Hurricane
1951: The Man in the Saddle
1952: Carson City; Springfield Rifle; Last of the Comanches
1953: House of Wax; The Stranger Wore a Gun; Thunder over the Plains
1954: Crime Wave; Riding Shotgun; Tanganyika; The Bounty Hunter
1955: The Indian Fighter
1957: Monkey on My Back; Hidden Fear
1959: The Two-Headed Spy (Brit.); Day of the Outlaw
1960: Man on a String
1961: Morgan the Pirate (Ital.)
1962: The Mongols
1964: Gold for the Caesars (Ital.)
1969: Play Dirty

RALPH DE VITO

1976: Death Collector

JOHN DEXTER (1935-) Brit.

(American films only)
1971: Pigeons

MAURY DEXTER (c. 1928-)

1960: The High-Powered Rifle; Walk Tall
1961: The Purple Hills; Womanhunt
1962: Air Patrol; The Firebrand; Young Guns of Texas
1963: The Day Mars Invaded Earth; House of the Damned; Police Nurse; Harbor Lights
1964: The Young Swingers; Surf Party
1965: Raiders from Beneath the Sea; The Naked Brigade; Wild on the Beach
1968: Maryjane; The Mini-Skirt Mob; The Young Animals; Born Wild
1970: Hell's Belles

DEANE H. DICKASON

1932: Virgins of Bali (doc.)

SAMUEL DIEGE (-1939)

1938: Water Rustlers; Singing Cowboy; Ride 'Em Cowboy
1939: Ride 'Em Cowgirl; The Singing Cowgirl

WILLIAM DIETERLE (1893-1972) Germ.

(American films only)
1931: Her Majesty Love; The Last Flight
1932: Jewel Robbery; Lawyer Man; Six Hours to Live; The Crash; Scarlet Dawn; Man Wanted
1933: The Devil's in Love; Grand Slam; Adorable; From Headquarters
1934: Fashions of 1934 (ret. Fashions); Fog over Frisco; Madame DuBarry; The Firebird
1935: Dr. Socrates; A Midsummer Night's Dream (w/ Max Reinhardt); Secret Bride; Concealment; Men on Her Mind
1936: Satan Met a Lady; The Story of Louis Pasteur; The White Angel

1937: Another Dawn; The Great O'Malley; The Life of Emile Zola
1938: Blockade
1939: The Hunchback of Notre Dame; Juarez
1940: A Dispatch from Reuters; Dr. Ehrlich's Magic Bullet
1941: All That Money Can Buy
1942: Syncopation; Tennessee Johnson
1944: I'll Be Seeing You; Kismet
1945: Love Letters; This Love of Ours
1946: The Searching Wind; Duel in the Sun (unc. w/ others)
1948: The Accused; Portrait of Jennie
1949: Rope of Sand; Volcano (Ital.)
1950: Dark City; Paid in Full; September Affair
1951: Peking Express; Red Mountain
1952: Boots Malone; The Turning Point
1953: Salome
1954: Elephant Walk
1956: Magic Fire
1957: Omar Khayyam
1964: Quick, Let's Get Married

JOHN FRANCIS DILLON (1887-1934)

1928: Scarlet Seas
1929: Children of the Ritz; Careers; Fast Life; Sally
1930: Bride of the Regiment; Spring Is Here; The Girl of the Golden West; Kismet; One Night at Susie's
1931: Millie; The Finger Points; The Reckless Hour; The Pagan Lady
1932: The Cohens and Kellys in Hollywood; Behind the Mask; Man About Town; Call Her Savage
1933: Humanity
1934: The Big Shakedown

CHARLES DILTZ

1932: Wild Women of Borneo

SCOTT DITTRICH

1976: Freewheelin'

DENVER DIXON (see VICTOR ADAMSON)

IVAN DIXON

1972: Trouble Man
1973: The Spook Who Sat by the Door

EDWARD DMYTRYK (1908-)

1935: The Hawk
1939: Television Spy
1940: Emergency Squad; Golden Gloves; Mystery Sea Raider; Her First Romance
1941: The Devil Commands; Under Age; Sweetheart of the Campus; The Blonde from Singapore; Secrets of the Lone Wolf; Confessions of Boston Blackie
1942: Counter-Espionage; Seven Miles from Alcatraz
1943: Hitler's Children; The Falcon Strikes Back; Captive Wild Woman; Behind the Rising Sun; Tender Comrade
1944: Murder, My Sweet
1945: Back to Bataan; Cornered
1946: Till the End of Time
1947: Crossfire; So Well Remembered (Brit.)
1949: The Hidden Room (Obsession-Brit.)
1950: Salt to the Devil (Give Us This Day-Brit.)
1952: Mutiny; The Sniper; Eight Iron Men
1953: The Juggler
1954: The Caine Mutiny; Broken Lance
1955: The End of the Affair (Brit.); Soldier of Fortune; The Left Hand of God
1956: The Mountain
1957: Raintree County
1958: The Young Lions; Warlock; The Blue Angel
1962: Walk on the Wild Side; The Reluctant Saint (Ital.)
1964: The Carpetbaggers; Where Love Has Gone
1965: Mirage
1966: Alvarez Kelly
1968: Anzio (Ital.); Shalako
1972: Bluebeard (Hung.)
1975: The Human Factor (Brit./Ital.)
1976: He Is My Brother

LARRY DOBKIN

1974: Like a Crow on a June Bug

LAWRENCE DOHENY

1961: Teenage Millionaire

VINCENT J. DONEHUE

1958: Lonelyhearts
1960: Sunrise at Campobello

STANLEY DONEN (1924-   )

1949: On the Town (w/ Gene Kelly)
1951: Royal Wedding
1952: Love Is Better Than Ever; Singin' in the Rain (w/ G. Kelly); Fearless Fagan
1953: Give a Girl a Break
1954: Deep in My Heart; Seven Brides for Seven Brothers
1955: It's Always Fair Weather (w/ G. Kelly)
1957: Funny Face; Kiss Them for Me; The Pajama Game (w/ G. Abbott)
1958: Indiscreet; Damn Yankees (w/ G. Abbott)
1960: The Grass Is Greener; Once More, With Feeling (Brit.); Surprise Package (Brit.)
1963: Charade
1966: Arabesque
1967: Bedazzled
1968: Two for the Road (Brit.)
1969: Staircase (Brit.)
1974: The Little Prince (Brit.)
1975: Lucky Lady
1978: Movie Movie

WALTER DONIGER (1917-   )

1954: Duffy of San Quentin; The Steel Cage
1956: The Steel Jungle
1958: Unwed Mother
1962: House of Women; Safe at Home!

CLIVE DONNER (1926-   ) Brit.

(American films only)
1965: What's New, Pussycat?
1967: Luv
1974: Old Dracula

RICHARD DONNER

1962: X-15
1968: Salt and Pepper (Brit.)
1973: Lola
1976: The Omen
1978: Superman

JACK DONOHUE (1908-   )

1948: Close-Up

1950: The Yellow Cab Man; Watch the Birdie
1954: Lucky Me
1961: Babes in Toyland
1965: Marriage on the Rocks
1966: Assault on a Queen

RON DORFMAN

1970: Groupies (doc. w/ Peter Nevard)

RUSSELL S. DOUGHTON, JR.

1967: The Hostage
1968: Fever Heat

GORDON DOUGLAS (1909-)

1936: General Spanky
1939: Zenobia
1940: Saps at Sea
1941: Road Show (w/ Hal Roach, Hal Roach, Jr.); Niagara Falls; Broadway Limited
1942: The Devil with Hitler; The Great Gildersleeve
1943: Gildersleeve's Bad Day; Gildersleeve on Broadway
1944: A Night of Adventure; Gildersleeve's Ghost; Girl Rush; The Falcon in Hollywood
1945: Zombies on Broadway; First Yank into Tokyo
1946: Dick Tracy vs. Cueball; San Quentin
1948: If You Knew Susie; The Black Arrow; Walk a Crooked Mile
1949: The Doolins of Oklahoma; Mr. Soft Touch (w/ Henry Levin)
1950: The Nevadan; The Fortunes of Captain Blood; Rogues of Sherwood Forest; Kiss Tomorrow Goodbye; Between Midnight and Dawn; The Great Missouri Raid
1951: Only the Valiant; I Was a Communist for the FBI; Come Fill the Cup
1952: Mara Maru; The Iron Mistress
1953: She's Back on Broadway; So This Is Love; The Charge at Feather River
1954: Them!; Young at Heart
1955: The McConnell Story; Sincerely Yours
1956: Santiago
1957: The Big Land; Bombers B-52
1958: Fort Dobbs; The Fiend Who Walked the West
1959: Up Periscope; Yellowstone Kelly
1961: Gold of the Seven Saints; The Sins of Rachel Cade; Claudelle Inglish
1962: Follow That Dream
1963: Call Me Bwana
1964: Robin and the Seven Hoods; Rio Conchos
1965: Sylvia; Harlow

1966: Stagecoach; Way ... Way Out
1967: In Like Flint; Chuka; Tony Rome
1968: The Detective; Lady in Cement
1970: Skullduggery; Barquero; They Call Me MISTER Tibbs!
1973: Slaughter's Big Rip-Off
1977: Viva Knievel!

KIRK DOUGLAS (1916-)

1973: Scalawag
1975: Posse

ROBERT DOWNEY (1936-)

1966: Chafed Elbows
1968: No More Excuses
1969: Putney Swope
1970: Pound
1972: Greaser's Palace
1976: Two Tons of Turquoise to Taos Tonight
1979: Jive

STAN DRAGOTI

1972: Dirty Little Billy
1979: Love at First Bite

OLIVER DRAKE

1934: Texas Tornado
1942: Today I Hang
1943: Border Buckaroos
1944: Trail of Terror
1946: Moon over Montana; Trail to Mexico; West of the Alamo; Ginger
1947: Deadline; Fighting Mustang; Shootin' Irons; Son of the Sierras; Sunset Carson Rides Again
1949: Across the Rio Grande; Brand of Fear; Lawless Code; Roaring Westward
1957: The Parson and the Law; A Lust to Kill

ARTHUR DREIFUSS (1908-)

1940: Mystery in Swing
1941: Reg'lar Fellers
1942: Baby Face Morgan; The Boss of Big Town; The Pay-Off
1943: Sarong Girl; Melody Parade; Campus Rhythm; Nearly Eighteen; The Sultan's Daughter

1944: Ever Since Venus
1945: Eadie Was a Lady; Boston Blackie Booked on Suspicion; Boston Blackie's Rendezvous; The Gay Senorita; Prison Ship
1946: Junior Prom; Freddie Steps Out; High School Hero
1947: Vacation Days; Betty Co-Ed; Little Miss Broadway; Two Blondes and a Redhead; Sweet Genevieve
1948: Glamour Girl; Mary Lou; I Surrender, Dear
1949: An Old-Fashioned Girl; Manhattan Angel; Shamrock Hill; There's a Girl in My Heart
1958: Life Begins at 17; The Last Blitzkrieg
1959: Juke Box Rhythm
1962: The Quare Fellow (Ire.)
1967: Riot on Sunset Strip; The Love-Ins
1968: For Singles Only; A Time to Sing; The Young Runaways

DONALD DRIVER

1973: The Naked Ape

CHARLES DUBIN

1957: Mister Rock and Roll
1976: Moving Violation

CLAUDE DU BOC

1975: One by One (doc.)

ARTHUR DUBS

1974: Vanishing Wilderness (doc. w/ Heinz Seilmann)

MICHAEL DUGAN

1976: Super Seal

DARYL DUKE

1973: Payday
1978: The Silent Partner (Can.)

SCOTT R. DUNLAP (1892-1970)

1929: One Stolen Night

PHILIP DUNNE (1908-)

1955: Prince of Players; The View from Pompey's Head
1956: Hilda Crane
1957: Three Brave Men
1958: In Love and War; Ten North Frederick
1959: Blue Denim
1961: Wild in the Country
1962: Lisa
1966: Blindfold

E. A. DUPONT (1891-1956) Germ.

(American films only)
1933: Ladies Must Love
1935: The Bishop Misbehaves
1936: Forgotten Faces; A Son Comes Home
1937: Night of Mystery; On Such a Night
1938: Love on Toast
1939: Hell's Kitchen (w/ Lewis R. Seiler)
1951: The Scarf; Pictura (w/ Luciano Emmer)
1953: Problem Girls; The Neanderthal Man; The Steel Lady
1954: Return to Treasure Island

RUDY DURAND

1979: Tilt

G. A. DURLAM

1932: Two-Fisted Justice

DICK DURRANCE

1951: Ski Champs

DAVID DURSTON

1970: I Drink Your Blood
1972: Stigma; Blue Sextet

LEON D'USSEAU

1932: The Girl from Calgary (w/ P. H. Whitman)

ROBERT DUVALL

1977: We're Not the Jet Set (doc.)

JULIEN DUVIVIER (1896-1967) Fr.

(American films only)
1938: The Great Waltz
1941: Lydia
1942: Tales of Manhattan
1943: Flesh and Fantasy
1944: The Imposter; Destiny (w/ R. LeBorg)

ALLAN DWAN (1885-)

1929: Frozen Justice; South Sea Rose; Tide of Empire; The Far Call; The Iron Mask
1930: What a Widow!
1931: Man to Man; Chances; Wicked
1932: While Paris Sleeps
1933: Her First Affair; Counsel's Opinion
1934: The Morning After; Hollywood Party
1935: Black Sheep; Beauty's Daughter (orig. Navy Wife)
1936: Song and Dance Man; Human Cargo; High Tension; 15 Maiden Lane
1937: Woman Wise; That I May Live; One Mile from Heaven; Heidi
1938: Rebecca of Sunnybrook Farm; Josette; Suez
1939: The Three Musketeers; The Gorilla; Frontier Marshal
1940: Sailor's Lady; Young People; Trail of the Vigilantes
1941: Rise and Shine; Look Who's Laughing
1942: Friendly Enemies; Here We Go Again
1943: Around the World
1944: Up in Mabel's Room; Abroad with Two Yanks
1945: Brewster's Millions
1946: Getting Gertie's Garter; Rendezvous with Annie
1947: Northwest Outpost; Calendar Girl; Driftwood
1948: The Inside Story; Angel in Exile (w/ P. Ford)
1949: The Sands of Iwo Jima
1950: Surrender
1951: Belle le Grand
1952: The Wild Blue Yonder; I Dream of Jeanie; Montana Belle
1953: The Woman They Almost Lynched; Sweethearts on Parade
1954: Flight Nurse; Silver Lode; Passion
1955: Cattle Queen of Montana; Escape to Burma; Pearl of the Pacific; Tennessee's Partner
1956: Slightly Scarlet; Hold Back the Night
1957: The River's Edge; The Restless Breed
1958: Enchanted Island
1961: The Most Dangerous Man Alive

KAYE H. DYAL

1974: Memory of Us

BOB DYLAN

1978: Renaldo and Clara

B. REEVES (BREEZY) EASON (1886-1956)

1929: Lariat Kid; Winged Horseman (w/ Arthur Rosson)
1930: Troopers Three (w/ Norman Taurog); The Roaring Ranch; Trigger Tricks; Spurs
1932: The Sunset Trail; Honor of the Press; The Heart Punch
1933: Cornered; Behind Jury Doors; Alimony Madness; Revenge at Monte Carlo; Her Resale Value; Dance Hall Hostess; Neighbors' Wives
1936: Red River Valley
1937: Land Beyond the Law; Empty Holsters; Prairie Thunder
1938: Sergeant Murphy; The Kid Comes Back; The Daredevil Drivers; Call of the Yukon (w/ John T. Coyle)
1939: Blue Montana Skies; Mountain Rhythm
1940: Men with Steel Faces (w/ Otto Brower)
1942: Murder in the Big House; Spy Ship
1943: Truck Busters; Murder on the Waterfront
1949: Rimfire

(Serials)

1931: The Galloping Ghost; The Vanishing Legion
1932: The Last of the Mohicans (w/ F. Beebe)
1934: The Law of the Wild (w/ A. Schaefer); Mystery Mountain (w/ O. Brower)
1935: The Phantom Empire (w/ O. Brower); The Adventures of Rex and Rinty (w/ Beebe); Fighting Marines (w/ J. Kane)
1936: Darkest Africa (w/ J. Kane); The Undersea Kingdom (w/ J. Kane)
1943: The Phantom
1945: Black Arrow
1947: The Desert Hawk
1949: King of the Jungleland (w/ J. Kane)

CHARLES EASTMAN

1973: The All-American Boy

GORDON EASTMAN

1976: North of the Sun (doc.); The Savage World

CLINT EASTWOOD (1930-)

1971: Play Misty for Me
1973: High Plains Drifter; Breezy
1975: The Eiger Sanction
1976: The Outlaw Josey Wales
1977: The Gauntlet

DREW EBERSON

1938: The Overland Express

DON EDMONDS

1974: Tender Loving Care

BLAKE EDWARDS (1922-)

1955: Bring Your Smile Along
1956: He Laughed Last
1957: Mr. Cory
1958: The Perfect Furlough; This Happy Feeling
1959: Operation Petticoat
1960: High Time
1961: Breakfast at Tiffany's
1962: Days of Wine and Roses; Experiment in Terror
1964: The Pink Panther; A Shot in the Dark
1965: The Great Race
1966: What Did You Do in the War, Daddy?
1967: Gunn
1968: The Party
1969: Darling Lili
1971: Wild Rovers
1972: The Carey Treatment
1974: The Tamarind Seed
1975: The Return of the Pink Panther (Brit.)
1976: The Pink Panther Strikes Again (Brit.)
1978: The Revenge of the Pink Panther
1979: 10

JAN EGLESON

1979: Billy in the Lowlands

RICHARD EINFELD

1957: Ghost Diver (w/ M. C. White)

ROBERT ELFSTROM

1969: Johnny Cash! (doc.)
1972: Pete Seeger ... A Song and a Stone
1973: The Gospel Road

CLYDE E. ELLIOT

1932: Bring 'Em Back Alive
1934: Devil Tiger; China Roars
1938: Booloo

GRACE ELLIOT

1931: The Three Racketeers; 10,000 and Broke; Him Who Has; Marriage à la Carte; The Spice of Life; Devil's Marriage; Splurge

ROBERT ELWYN

1953: That Man from Tangier

ROBERT EMMETT (see ROBERT E. TANSEY)

ED EMSHWILLER

1970: Image, Flesh and Voice
1971: Branches

ROBERT A. ENDELSON

1979: Staying Alive

CYRIL ENDFIELD (1914-   ) S. Afr.

(American films only)
1946: Gentleman Joe Palooka
1947: Stork Bites Man
1948: The Argyle Secrets
1949: Joe Palooka in the Big Fight
1950: The Sound of Fury (ret. Try and Get Me)
1952: Tarzan's Savage Fury
1961: Mysterious Island
1964: Hide and Seek
1969: De Sade

MORRIS ENGEL (1918-)

1956: Lovers and Lollipops (w/ Ruth Orkin)
1960: Weddings and Babies

JOHN ENGLISH (1903-1969)

1935: His Fighting Blood; Red Blood of Courage
1937: Arizona Days; Whistling Bullets
1938: Call the Mesquiteers
1941: Gangs of Sonora
1942: Code of the Outlaw; The Yukon Patrol; Valley of Hunted Men; Westward Ho; Phantom Plainsmen; Shadows on the Range; Raiders of the Range
1943: Death Valley Manhunt; The Man from Thunder River; Overland Mail Robbery; Dead Man's Gulch; The Black Hills Express; Thundering Trails
1944: Grissly's Millions; San Fernando Valley; The Port of Missing Thieves; Call of the South Seas; The Laramie Trail; Faces in the Fog; Silver City Kid
1945: Utah; Don't Fence Me In; The Phantom Speaks; Behind City Lights
1946: Murder in the Music Hall
1947: Trail to San Antone; The Last Round-Up
1948: The Strawberry Roan
1949: The Cowboy and the Indians; Loaded Pistols; Riders of the Sky; Riders of the Whistling Pines; Rim of the Canyon
1950: Beyond the Purple Hills; The Blazing Sun; Cow Town; Indian Territory; Mule Train; Sons of New Mexico; The Whipped
1951: Gene Autry and the Mounties; Hills of Utah; Silver Canyon; Valley of Fire; Whirlwind

(Serials)

1937: Zorro Rides Again (w/ W. Witney)
1938: Dick Tracy Returns (w/ W. Witney); Fighting Devil Dogs (w/ W. Witney and R. Becher); Hawk of the Wilderness (w/ W. Witney); The Lone Ranger (w/ W. Witney)
1939: Daredevils of the Red Circle; Dick Tracy's G-Men; The Lone Ranger Rides Again; Zorro's Fighting Legion (all w/ W. Witney)
1940: Drums of Fu Manchu; Adventures of Red Ryder; Hi Yo Silver; King of the Royal Mounted; Mysterious Dr. Satan (all w/ W. Witney)
1941: Adventures of Capt. Marvel; Dick Tracy vs. Crime, Inc.; Jungle Girl; King of the Texas Rangers (all w/ W. Witney)
1942: Perils of Nyoka; Spy Smasher (both w/ W. Witney)
1943: Daredevils of the West
1944: Captain America (w/ E. Clifton)
1952: Dick Tracy vs. Phantom Empire (w/ Witney)

GEORGE H. ENGLUND (1926-)

1963: The Ugly American
1964: Signpost to Murder
1971: Zachariah
1972: Snow Job

RAY (RAYMOND E.) ENRIGHT (1896-1965)

1929: Little Wildcat; Stolen Kisses; Kid Gloves; Skin Deep
1930: Song of the West; Golden Dawn; Dancing Sweeties; Scarlet Pages
1932: Play Girl; The Tenderfoot
1933: Blondie Johnson; The Silk Express; Tomorrow at Seven; Havana Widows
1934: I've Got Your Number; Twenty Million Sweethearts; The Circus Clown; Dames; The St. Louis Kid
1935: While the Patient Slept; Traveling Saleslady; Alibi Ike; We're in the Money; Miss Pacific Fleet; Sing Me a Love Song
1936: Snowed Under; Earthworm Tractors; China Clipper
1937: Ready, Willing, and Able; Slim, the Singing Marine; Back in Circulation
1938: Swing Your Lady; Gold Diggers in Paris; Hard to Get; Going Places
1939: Naughty but Nice; Angels Wash Their Faces; On Your Toes
1940: Brother Rat and a Baby; An Angel from Texas; The River's End
1941: The Wagons Roll at Night; Thieves Fall Out; Bad Men of Missouri; Law of the Tropics; Wild Bill Hickok Rides
1942: The Spoilers; Men of Texas; Sin Town
1943: Good Luck, Mr. Yates; The Iron Major; Gung Ho!
1945: China Sky; Man Alive
1946: One Way to Love
1947: Trail Street
1948: Albuquerque; Return of the Bad Men; Coroner Creek
1949: South of St. Louis
1950: Montana; Kansas Raiders
1951: Flaming Feather
1953: The Man from Cairo

HENRY EPHRON (1912-)

1958: Sing Boy Sing

JEROME EPSTEIN

1969: The Adding Machine

RICHARD ERDMAN (1925-)

1971: Bleep
1973: The Brothers O'Toole

A. F. ERICKSON

1930: Rough Romance; The Lone Star Ranger; Under Suspicion
1932: This Sporting Age (w/ A. W. Bennison)

DAN ERIKSEN

1966: A Midsummer Night's Dream

JOHN ERMAN

1971: Making It
1973: Ace Eli and Roger of the Sky

CHESTER ERSKINE (1905-)

1934: Midnight
1935: Frankie and Johnnie
1947: The Egg and I
1949: Take One False Step
1952: Androcles and the Lion; A Girl in Every Port
1972: Irish Whiskey Rebellion

DWAIN ESPER

1934: Maniac

HARRY ESSEX (1910-)

1953: I, the Jury
1955: Mad at the World

HOWARD ESTABROOK (1884-1978)

1944: Heavenly Days

ALEXANDER ESWAY

1931: Shadows
1945: Steppin' in Society

JOHN EVANS

1974: Black Godfather

JASON EVERS

1963: The Brain That Wouldn't Die

DOUGLAS FAIRBANKS (1883-1939)

1931: Around the World in Eighty Minutes with Douglas Fairbanks (w/ Victor Fleming)

WILLIAM FAIRCHILD

1957: John and Julie
1958: The Silent Enemy

JAMAA FANAKA

1975: Welcome Home, Brother Charles
1979: Penitentiary

JAMES FARGO

1976: The Enforcer
1978: Caravans; Every Which Way but Loose

JOHN FARRIS

1975: Dear Dead Delilah

JOHN VILLIERS FARROW (1904-1963)

1937: Men in Exile; West of Shanghai (War Lord); The Invisible Menace
1938: She Loved a Fireman; Little Miss Thoroughbred; My Bill; Broadway Musketeers
1939: Women in the Wind; The Saint Strikes Back; Sorority House; Five Came Back; Full Confession; Reno
1940: Married and in Love; A Bill of Divorcement
1942: Wake Island; Commandos Strike at Dawn
1943: China
1944: The Hitler Gang
1945: You Came Along
1946: Two Years Before the Mast; California

1947: Easy Come, Easy Go; Blaze of Noon; Calcutta
1948: The Big Clock; The Night Has a Thousand Eyes; Beyond Glory
1949: Alias Nick Beal
1951: His Kind of Woman; Submarine Command
1953: Ride, Vaquero; Plunder of the Sun; Botany Bay; Hondo
1954: A Bullet Is Waiting
1955: The Sea Chase
1956: Back from Eternity
1957: The Unholy Wife
1959: John Paul Jones

WILLIAM FEIGENBAUM

1976: Hugo the Hippo (anim.)

BARRY FEINSTEIN

1968: You Are What You Eat

FELIX E. FEIST, JR. (1906-1965)

1933: The Deluge
1943: All by Myself; You're a Lucky Fellow, Mr. Smith
1944: This Is the Life; Pardon My Rhythm; Reckless Age
1945: George White's Scandals
1947: The Devil Thumbs a Ride
1948: The Winner's Circle
1949: The Threat; Guilty of Treason
1950: The Golden Gloves Story; The Man Who Cheated Himself
1951: Tomorrow Is Another Day; The Basketball Fix
1952: This Woman Is Dangerous; The Big Trees; The Man Behind the Gun
1953: Donovan's Brain
1955: Pirates of Tripoli

PAUL FEJOS (1897-1963) Hung.

(American films only)
1929: Broadway; The Last Performance
1930: Captain of the Guard (completed by J. S. Robertson)

MARTY FELDMAN (1938-) Brit.

1977: The Last Remake of Beau Geste
1979: In God We Trust

SEYMOUR FELIX (1892-1961)

1931: Girls Demand Excitement
1932: Stepping Sisters

GEORG FENADY

1973: Arnold; Terror in the Wax Museum

LESLIE FENTON (1902-1978)

1939: Tell No Tales; Stronger Than Desire
1940: The Man from Dakota; The Golden Fleecing
1941: The Saint's Vacation (Brit.)
1943: There's a Future in It (Brit.)
1944: Tomorrow the World!
1946: Pardon My Past
1948: On Our Merry Way (w/ King Vidor); Saigon; Lulu Belle; Whispering Smith
1949: Streets of Laredo
1950: The Redhead and the Cowboy

NORMAN FERGUSON

1945: The Three Caballeros

JOSE FERRER (1921-)

1955: The Shrike
1956: The Cockleshell Heroes (Brit.); The Great Man
1958: I Accuse; The High Cost of Loving
1961: Return to Peyton Place
1962: State Fair

MEL FERRER (MELCHIOR G. FERRER) (1917-)

1945: The Girl of the Limberlost
1950: The Secret Fury; Vendetta (w/ Max Ophuls, Preston Sturges, Stuart Heisler, Howard Hughes; signed by Ferrer)
1959: Green Mansions
1966: Every Day Is a Holiday (Span.)

FRANK FERRIN

1955: Sabaka

E. I. FESSLER

1956: Bayou (w/ H. Daniels)

JACQUES FEYDER (1885-1948) Belg.

1930: Anna Christie (Germ. version)
1931: Daybreak; Son of India

JULIUS D. FIEGELSON

1971: The Windsplitter

LEONARD FIELDS

1934: Manhattan Love Song; King Kelly of the U.S.A.
1935: Streamline Express

MICHAEL FINDLAY

1976: Shriek of the Mutilated

SEATON FINDLAY

1975: Janis (doc. w/ H. Alk)

MICHAEL FINK

1975: Force Four
1976: Velvet Smooth

MICHAEL J. FINN

1974: The Black Generation

EDWARD FINNEY

1941: Silver Stallion; Riot Squad
1942: King of the Stallions
1947: Queen of the Amazons

CINDA FIRESTONE

1974: Attica (doc.)

GEORGE FITZMAURICE (1885-1940)

1929: His Captive Woman; Man and the Moment; The Locked Door; Tiger Rose
1930: The Bad One; One Heavenly Night; The Devil to Pay
1931: Strangers May Kiss; The Unholy Garden
1932: Mata Hari; As You Desire Me
1934: All Men Are Enemies
1936: Petticoat Fever; Suzy
1937: The Last of Mrs. Cheyney (begun and signed by Richard Boleslawski); The Emperor's Candlesticks; Live, Love and Learn
1938: Arsene Lupin Returns; Vacation from Love
1940: Adventure in Diamonds

JAMES FITZPATRICK

1930: The Lady of the Lake
1945: Song of Mexico

ROBERT S. FIVESON

1979: The Clonus Horror

ROBERT FLAHERTY (1884-1951)

1931: Tabu
1934: Man of Aran (doc. Brit.)
1937: Elephant Boy (Brit.)
1942: The Land (doc.)
1948: Louisiana Story (doc.)

DAVE FLEISCHER (1894-1979)

1939: Gulliver's Travels (anim.)
1941: Mr. Bug Goes to Town (anim.)

RICHARD O. FLEISCHER (1916-)

1946: Child of Divorce
1947: Banjo
1948: So This Is New York; Bodyguard
1949: The Clay Pigeon; Trapped; Follow Me Quietly; Make Mine Laughs
1950: Armored Car Robbery
1952: The Narrow Margin; The Happy Time
1953: Arena
1954: 20,000 Leagues Under the Sea

1955: Violent Saturday; The Girl in the Red Velvet Swing
1956: Bandido; Between Heaven and Hell
1958: The Vikings
1959: These Thousand Hills; Compulsion
1960: Crack in the Mirror
1961: The Big Gamble
1962: Barabbas (Ital.)
1966: Fantastic Voyage
1967: Doctor Dolittle (Brit.)
1968: The Boston Strangler
1969: Che!
1970: Tora! Tora! Tora! (w/ Toshio Masuda, Kinji Fukusaku)
1971: Ten Rillington Place (Brit.); The Last Run (Span.); See No Evil (Brit.)
1972: The New Centurions
1973: Soylent Green; The Don Is Dead
1974: The Spikes Gang (Span.); Mr. Majestyk
1975: Mandingo
1976: The Incredible Sarah (Brit.)
1978: Crossed Swords (Brit.)
1979: Ashanti

VICTOR FLEMING (1883-1949)

1929: The Virginian; Wolf Song
1930: Renegades; Common Clay
1931: Around the World in 80 Minutes (w/ D. Fairbanks)
1932: Red Dust; The Wet Parade
1933: Bombshell (ret. Blonde Bombshell); The White Sister
1934: Treasure Island
1935: The Farmer Takes a Wife; Reckless
1937: Captains Courageous
1938: Test Pilot
1939: Gone with the Wind; The Wizard of Oz
1941: Dr. Jekyll and Mr. Hyde
1942: Tortilla Flat
1943: A Guy Named Joe
1945: Adventure
1948: Joan of Arc

GORDON FLEMYNG

1963: Just for Fun!
1968: The Split; Great Catherine

THEODORE J. FLICKER (1930-)

1964: The Troublemaker
1967: The President's Analyst
1970: Up in the Cellar

JAMES T. FLOCKER

1976: Secret of Navajo Cave

JAMES FLOOD (1895-1953)

1929: Midstream; Mister Antonio
1930: Sisters; Swellhead
1931: She-Wolf; Mother's Millions
1932: Life Begins (w/ E. Nugent); Under-Cover Man; The Mouthpiece (w/ E. Nugent)
1934: All of Me; Such Women Are Dangerous; Wings in the Dark
1935: Shanghai
1936: We're Only Human; Everybody's Old Man
1937: Midnight Madonna; Scotland Yard Commands
1939: Off the Record
1947: The Big Fix; Stepchild

JOHN FLOREA (1916-)

1972: Pickup on 101

ROBERT FLOREY (1900-)

1929: The Hole in the Wall; The Coconuts (w/ Joseph Stanley); Battle of Paris
1932: Murders in the Rue Morgue; The Man Called Back; Those We Love
1933: Girl Missing; Ex-Lady; The House on 56th Street
1934: Bedside; Smarty; Registered Nurse; I Sell Anything
1935: I Am a Thief; The Woman in Red; The Florentine Dagger; Don't Bet on Blondes; Going Highbrow; Ship Cafe; The Pay-Off
1936: The Preview Murder Mystery; 'Til We Meet Again; Hollywood Boulevard
1937: Outcast; King of Gamblers; Mountain Music; This Way Please; Daughter of Shanghai
1938: Dangerous to Know; King of Alcatraz
1939: Disbarred; Hotel Imperial; The Magnificent Fraud; Death of a Champion
1940: Women Without Names; Parole Fixer
1941: The Face Behind the Mask; Meet Boston Blackie; Two in a Taxi; Dangerously They Live
1942: Lady Gangster (as "Florian Roberts")
1943: The Desert Song
1944: Man from Frisco; Roger Touhy, Gangster
1945: God Is My Co-Pilot; Danger Signal
1946: The Beast with Five Fingers
1948: Tarzan and the Mermaids; Rogues Regiment
1949: Outpost in Morocco; The Crooked Way
1950: The Vicious Years; Johnny One-Eye

EMMETT J. FLYNN

1929: Hold Your Man; The Shannons of Broadway

JOHN FLYNN

1968: The Sergeant
1972: The Jerusalem File (Isr./U.S.)
1973: The Outfit
1977: Rolling Thunder
1979: Defiance

PETER FONDA (1939-)

1971: The Hired Hand
1975: Idaho Transfer
1979: Wanda Nevada

JORGE FONS

1973: Jory

RICHARD FONTAINE

1975: The Sins of Rachel

BRYAN FORBES (1926-) Brit.

(American films only)
1965: King Rat
1975: The Stepford Wives

JOHN FORD (1895-1973)

1929: Salute (w/ D. Butler); Strong Boy; The Black Watch
1930: Men Without Women; Born Reckless; Up the River
1931: The Brat; Arrowsmith; The Seas Beneath
1932: Air Mail; Flesh
1933: Dr. Bull; Pilgrimage
1934: The Lost Patrol; The World Moves On; Judge Priest
1935: The Informer; Steamboat 'Round the Bend; The Whole Town's Talking
1936: Mary of Scotland; The Plough and the Stars; The Prisoner of Shark Island
1937: The Hurricane; Wee Willie Winkie
1938: Four Men and a Prayer; Submarine Patrol
1939: Drums Along the Mohawk; Stagecoach; Young Mr. Lincoln

1940: The Grapes of Wrath; The Long Voyage Home
1941: How Green Was My Valley; Tobacco Road
1945: They Were Expendable
1946: My Darling Clementine
1947: The Fugitive
1948: Fort Apache; Three Godfathers
1949: She Wore a Yellow Ribbon
1950: Rio Grande; Wagonmaster; When Willie Comes Marching Home
1952: The Quiet Man; What Price Glory
1953: Mogambo; The Sun Shines Bright
1955: The Long Gray Line; Mr. Roberts
1956: The Searchers
1957: The Rising of the Moon (Ire.); The Wings of Eagles
1958: The Last Hurrah
1959: The Horse Soldiers; Gideon of Scotland Yard
1960: Sergeant Rutledge
1961: Two Rode Together
1962: The Man Who Shot Liberty Valance
1963: Donovan's Reef; How the West Was Won (w/ others)
1964: Cheyenne Autumn
1966: Seven Women

PHILIP FORD

1945: The Tiger Woman
1946: Valley of the Zombies; The Last Crooked Mile; Crime of the Century; The Invisible Informer; The Inner Circle; The Mysterious Mr. Valentine
1947: The Wild Frontier; Bandits of Dark Canyon; The Web of Danger
1948: Angel in Exile (w/ P. Ford); The Bold Frontiersman; The Denver Kid; California Firebrand; Desperadoes of Dodge City; Marshal of Amarillo; The Timber Trail; Train to Alcatraz (w/ G. Geraghty)
1949: Law of the Golden West; The Wyoming Bandit; Hideout; Outcasts of the Trail; Pioneer Marshal; Powder River; Prince of the Plains; Ranger of Cherokee Strip; San Antone Ambush; South of Rio
1950: Buckaroo Sheriff of Texas; The Old Frontier; Prisoners in Petticoats; Redwood Forest Trail; Trial Without Jury; The Vanishing Westerner
1951: Wells Fargo Gunmaster; The Dakota Kid; Missing Women; Pride of Maryland; Rodeo King and the Senorita; Utah Wagon Train
1952: Bal Tabarin; Desperadoes' Outpost

WESLEY FORD

1933: Her Forgotton Past; Secret Sinners
1935: $20 a Week

EUGENE J. FORDE (1898-)

1929: Outlawed; Big Diamond Robbery
1933: Smoky
1934: Charlie Chan in London
1935: Mystery Woman; The Great Hotel Murder; Your Uncle Dudley
1936: The Country Beyond; 36 Hours to Kill
1937: Midnight Taxi; Step Lively, Jeeves!; The Lady Escapes; Charlie Chan on Broadway; Charlie Chan at Monte Carlo
1938: International Settlement; One Wild Night; Meet the Girls
1939: Inspector Hornleigh (Brit.); The Honeymoon's Over
1940: Charlie Chan's Murder Cruise; Pier 13; Michael Shayne, Private Detective; Charter Pilot
1941: Sleepers West; Dressed to Kill; Buy Me That Town; Man at Large
1942: Right to the Heart; Berlin Correspondent
1943: The Crime Doctor's Strangest Case
1944: Shadows in the Night
1947: Backlash; Jewels of Brandenburg; The Crimson Key; The Invisible Wall

CARL FOREMAN (1914-)

1963: The Victors

MICHAEL FORLONG

1961: The Green Helmet

MILOS FORMAN (1932-) Czech.

(American films only)
1971: Taking Off
1973: Visions of Eight (w/ others)
1975: One Flew over the Cuckoo's Nest
1979: Hair

ED FORSYTH

1973: Superchick

BOB FOSSE (1927-)

1969: Sweet Charity
1972: Cabaret
1974: Lenny
1979: All That Jazz

HARRY FOSTER (1906-)

1958: Let's Rock!

HARVE FOSTER

1946: Song of the South
1947: The Fabulous Joe (w/ Bernard Carr)

LEWIS R. FOSTER (1906-)

1936: Love Letters of a Star (w/ Milton Carruth)
1937: She's Dangerous (w/ Milton Carruth); Armored Car; The Man Who Cried Wolf
1949: The Lucky Stiff; El Paso; Manhandled; Captain China
1950: The Eagle and the Hawk
1951: The Last Outpost; Passage West; Crosswinds; Hong Kong
1953: Tropic Zone; Jamaica Run; Those Redheads from Seattle
1955: Crashout; Top of the World
1956: The Bold and the Brave; Dakota Incident
1958: Tonka
1960: The Sign of Zorro (w/ Norman Foster)

NORMAN FOSTER (1900-1976)

1936: I Cover Chinatown
1937: Fair Warning; Think Fast, Mr. Moto; Thank You, Mr. Moto
1938: Walking Down Broadway; Mysterious Mr. Moto; Mr. Moto Takes a Chance
1939: Mr. Moto's Last Warning; Charlie Chan in Reno; Mr. Moto Takes a Vacation; Charlie Chan at Treasure Island
1940: Charlie Chan in Panama; Viva Cisco Kid
1941: Ride, Kelly, Ride; Scotland Yard
1942: Journey into Fear (w/ Orson Welles)
1948: Rachel and the Stranger; Kiss the Blood off My Hands
1949: Tell It to the Judge
1950: Father Is a Bachelor (w/ Abby Berlin); Woman on the Run
1952: Navajo; Sky Full of Moon
1953: Sombrero
1955: Davy Crockett, King of the Wild Frontier
1960: The Sign of Zorro (w/ Lewis R. Foster)
1966: Indian Paint
1967: Brighty of the Grand Canyon

GENE FOWLER, JR.

1957: I Was a Teenage Werewolf
1958: Gang War; Showdown at Boot Hill; I Married a Monster from Outer Space
1959: Here Come the Jets; The Rebel Set; The Oregon Trail

DOUGLAS FOWLEY (1911-)

1960: Macumba Love

WALLACE FOX (1895-1958)

1929: Come and Get It; The Amazing Vagabond; Laughing at Death
1931: Partners of the Trail
1932: The Cannonball Express; Devil on Deck
1935: Red Morning; Powdersmoke Range
1936: Yellow Dust; The Last of the Mohicans (w/ George B. Seitz)
1937: Racing Lady
1938: The Mexicali Kid; The Gun Packer
1941: The Lone Star Vigilantes; Bowery Blitzkrieg
1942: Bullets for Bandits; The Corpse Vanishes; Let's Get Tough!; Smart Alecks; Bowery at Midnight; 'Neath Brooklyn Bridge
1943: Kid Dynamite; The Ghost Rider; Outlaws of Stampede Pass; The Girl from Monterey; Career Girl
1944: Men on Her Mind; The Great Mike; Riders of the Santa Fe; Pride of the Plains; Block Busters; Song of the Range; Million Dollar Kid
1945: Mr. Muggs Rides Again; Bad Men of the Border; Code of the Lawless; Trail to Vengeance; Pillow of Death
1946: Gun Town; Rustler's Round-Up; Wild Beauty; Lawless Breed; Gunman's Code
1948: Docks of New York; The Valiant Hombre
1949: The Gay Amigo; The Daring Caballero; Western Renegades
1950: Fence Riders; West of Wyoming; Over the Border; Gunslingers; Six-Gun Mesa; Arizona Territory; Silver Raiders; Outlaw Gold
1951: Montana Desperado; Blazing Bullets

(Serials)

1945: Brenda Starr, Reporter
1947: Jack Armstrong; The Vigilante

BRYAN FOY (1896-1977)

1928: The Home Towners; Lights of New York
1929: Queen of the Nightclubs; The Royal Box
1931: The Gorilla

WILLIAM FRAKER (1923-)

1970: Monte Walsh
1973: A Reflection of Fear

JESS FRANCO

1971: Venus in Furs

CHARLES FRANK

1951: The Inheritance

ERNEST L. FRANK

1933: Nagana
1934: One Exciting Adventure

MELVIN FRANK (c. 1917-)

(In collaboration with Norman Panama)
1950: The Reformer and the Redhead
1951: Strictly Dishonorable; Callaway Went Thataway
1952: Above and Beyond
1954: Knock on Wood
1956: The Court Jester; That Certain Feeling; Alone
1959: The Jayhawkers; Li'l Abner
1960: The Facts of Life
1964: Strange Bedfellows
1968: Buona Sera, Mrs. Campbell
1973: A Touch of Class (Brit./Span.)
1975: The Prisoner of Second Avenue
1976: The Duchess and the Dirtwater Fox
1979: Lost and Found

T. C. FRANK (see TOM LAUGHLIN)

JOHN FRANKENHEIMER (1930-)

1957: The Young Stranger
1961: The Young Savages
1962: All Fall Down; Birdman of Alcatraz; The Manchurian Candidate
1964: Seven Days in May
1965: The Train
1966: Grand Prix; Seconds
1968: The Fixer
1969: The Extraordinary Seaman; The Gypsy Moths
1970: I Walk the Line
1971: The Horsemen
1973: The Iceman Cometh
1974: 99 and 44/100% Dead

1975: The French Connection II
1977: Black Sunday
1979: Prophecy

CHESTER M. FRANKLIN (1890-1948)

1928: Detectives
1932: File No. 113; The Stoker; Vanity Fair; A Parisian Romance
1933: The Iron Master
1935: Sequoia
1936: Tough Guy

SIDNEY FRANKLIN (1893-1972)

1928: The Actress
1929: The Last of Mrs. Cheyney; Devil May Care; Wild Orchids
1930: A Lady's Morals; The Lady of Scandal; Soul Kiss
1931: The Guardsman; Private Lives
1932: Smilin' Through
1933: Reunion in Vienna
1934: The Barretts of Wimpole Street
1935: The Dark Angel
1937: The Good Earth (w/ others)
1946: Duel in the Sun (unc. w/ others)
1957: The Barretts of Wimpole Street

WENDELL JAMES FRANKLIN

1971: The Bus Is Coming

HARRY L. FRASER

1931: Montana Kid; Oklahoma Jim
1932: From Broadway to Cheyenne; Land of Wanted Men; Ghost City; The Reckoning; Texas Pioneers; Mason of the Mounted; Law of the North; Honor of the Mounted; The Savage Girl
1933: The Fighting Parson; The Fugitive; Diamond Trail; Rainbow Ranch
1934: Fighting Through; 'Neath Arizona Skies; Randy Rides Alone
1935: Rustlers' Paradise; Wagon Trail; Wild Mustang; Fighting Pioneers; Saddle Aces; Last of the Clintons
1936: Aces Wild; Feud of the West; Ghost Town; The Riding Avenger; Hair-Trigger Casey; Cavalcade of the West; Romance Rides the Range
1937: Heroes of the Alamo; Spirit of Youth; Galloping Dynamite
1938: Six Shootin' Sheriff; Fury Below; Songs and Saddles
1940: Lightning Strikes West; The Phantom Rancher

1941: Jungle Man
1943: The Old Chisholm Trail
1944: Brand of the Devil; Gunsmoke Mesa; Outlaw Roundup
1945: Enemy of the Law; Flaming Bullets; Frontier Fugitives; Navajo Kid; Three on the Saddle
1946: Ambush Trail; Six-Gun Man; Thunder Town; Six Gun for Hire
1947: The White Gorilla
1949: Stallion Canyon

(Serials)

1933: Wolf Dog (w/ C. Clark)
1937: Jungle Menace (w/ G. Melford)

JAMES FRAWLEY

1971: The Christian Licorice Store
1973: Kid Blue
1976: The Big Bus
1979: The Muppet Movie

STEPHEN FREARS

1972: Gumshoe

JEROLD FREEDMAN

1972: Kansas City Bomber

JOEL F. FREEDMAN

1970: Skezag (doc. w/ Philip F. Messina)

THORNTON FREELAND (1898-)

1929: Three Live Ghosts
1930: Be Yourself; Whoopee
1931: Six Cylinder Love; The Secret Witness; Terror by Night
1932: They Call It Sin; Week-end Marriage; The Unexpected Father; Love Affair
1933: Flying Down to Rio
1934: George White's Scandals
1935: Skylarks; Brewster's Millions
1936: The Amateur Gentleman; Accused
1937: The Gaiety Girls (Brit.); Dark Sands (Brit.); Over the Moon (w/ W. K. Howard)
1940: So This Is London

1941: Marry the Boss's Daughter; Too Many Blondes
1947: The Brass Monkey (alt. Lucky Mascot)
1948: Meet Me at Dawn
1949: Dear Mr. Prohack Pentagon

AL FREEMAN, JR.

1971: A Fable

JOEL R. FREEMAN

1975: Broken Treaty at Battle Mountain (doc.)

HOWARD FREEN

1974: Dirty O'Neil (w/ Lewis Teague)

HUGO FREGONESE (1908-) Arg.

1943: Pampa Barbara (Arg./w/ Lucas Demare)
1946: Where Words Fail (Arg.)
1950: One Way Street; Saddle Tramp
1951: Apache Drums; Mark of the Renegade
1952: My Six Convicts; Untamed Frontier
1953: Blowing Wild; Decameron Nights
1954: The Man in the Attic; The Raid; Black Tuesday
1958: Harry Black and the Tiger (Brit.)
1959: The Beast of Marseilles (Brit; Seven Thunders)
1962: Marco Polo (Ital.)
1967: Shatterhand! (Germ.); Savage Pampas (Span.)

EUGENE FRENKE

1934: Girl in the Cage; Life Returns
1954: Miss Robinson Crusoe

WILLIAM FRESHMAN

1948: The Plot to Kill Roosevelt

JAY FREUND

1979: The American Game

KARL FREUND (1890-1969) Czech.

1932: The Mummy
1933: Moonlight and Pretzels
1934: Madame Spy; Countess of Monte Cristo; Uncertain Lady; I Give My Love; Gift of Gab
1935: Mad Love

RICHARD FRIEDENBERG

1974: The Life and Times of Grizzly Adams
1976: The Adventures of Frontier Fremont
1979: The Bermuda Triangle

DAVID FRIEDKIN

1956: Hot Summer Night
1958: Handle With Care

WILLIAM FRIEDKIN (1939-)

1967: Good Times
1968: The Night They Raided Minsky's; The Birthday Party
1970: The Boys in the Band
1971: The French Connection
1973: The Exorcist
1977: Sorcerer
1978: The Brink's Job

DAVID FRIEDMAN

1938: Convicts at Large (w/ S. Beal)

SEYMOUR FRIEDMAN (1917-)

1948: Trapped by Boston Blackie
1949: Boston Blackie's Chinese Venture; Crime Doctor's Diary; The Devil's Henchman; Prison Warden; Chinatown at Midnight; Rusty's Birthday; Rusty Saves a Life
1950: Bodyhold; Customs Agent; Rookie Fireman; Counterspy Meets Scotland Yard
1951: Criminal Lawyer; Her First Romance; The Son of Dr. Jekyll
1952: Loan Shark
1953: Flame of Calcutta; I'll Get You (Brit.)
1954: Khyber Patrol; The Saint's Girl Friday (Brit.)
1955: African Manhunt
1956: Secret of Treasure Mountain

ROBERT FRIEND

1970: Tarzan's Deadly Silence

LARRY FRISCH (1929-)

1956: Tel Aviv Taxi (Isr.)
1962: Pillar of Fire (Isr.)
1963: Casablan (Gr.)

GUNTHER FRITCH

1944: The Curse of the Cat People (w/ Robert Wise)
1947: Cigarette Girl

LEE FROST

1971: Chrome and Hot Leather; The Scavengers
1972: The Thing with Two Heads
1974: Policewomen
1975: The Black Gestapo
1976: Dixie Dynamite

ROBERT FUEST (1927-) Brit.

(American films only)
1975: The Devil's Rain

CHARLES FUHR

1943: Bomber's Moon

LESTER FULLER

1944: You Can't Ration Love

SAMUEL FULLER (1911-)

1949: I Shot Jesse James
1950: The Baron of Arizona
1951: Fixed Bayonets; The Steel Helmet
1952: Park Row
1953: Pickup on South Street
1954: Hell and High Water
1955: House of Bamboo
1956: Run of the Arrow

1957: China Gate; Forty Guns
1958: Verboten!
1959: Crimson Kimono
1961: Underworld, U.S.A.
1962: Merrill's Marauders
1963: Shock Corridor
1964: The Naked Kiss
1968: Caine
1969: Shark!
1972: Dead Pigeon on Beethoven Street

ALLEN FUNT

1970: What Do You Say to a Naked Lady?
1972: Money Talks (doc.)

SIDNEY J. FURIE (1933-   ) Can.

(American films only)
1966: The Appaloosa
1967: The Naked Runner (Brit.)
1969: The Lawyer
1970: Little Fauss and Big Halsy
1972: Lady Sings the Blues
1973: Hit!
1975: Sheila Levine Is Dead and Living in New York City
1976: Gable and Lombard
1978: The Boys in Company C.
1979: Night of the Juggler

MARTIN GABEL (1912-   )

1947: The Lost Moment

ALAN GADNEY

1974: Moon Child

ROBERT GAFFNEY (1931-   )

1965: Frankenstein Meets the Space Monster

GEORGE GAGE

1977: Skateboard

JOHN GAGE

1948: The Velvet Touch

GEORGE GALE

1975: Mysteries from Beyond the Earth (doc.)

TIMOTHY GALFAS

1979: Sunnyside

DONALD GALLAHER

1929: Nix on Dames

PAUL GALLICO

1960: Next to No Time

ALBERT C. GANNAWAY (1920-   )

1956: Hidden Guns; Daniel Boone--Trail Blazer (w/ Ismael Rodriguez)
1957: The Badge of Marshal Brennan; Raiders of Old California
1958: Man or Gun; No Place to Land
1959: Plunderers of Painted Flats

ALVIN GANZER

1953: The Girls of Pleasure Island (w/ F. H. Herbert)
1956: The Leather Saint
1958: Country Music Holiday
1965: When the Boys Meet the Girls
1967: Three Bites of the Apple

CYRIL GARDNER

1930: Only Saps Work (w/ E. H. Knopf); The Royal Family of Broadway (w/ G. Cukor)
1931: Reckless Living
1932: Doomed Battalion

ROBERT GARDNER

1974: Rivers of Sand (doc.)

LEO GAREN

1973: Hex

JACK GARFEIN (1930-) Czech

1957: The Strange One
1961: Something Wild

MIKE GARGIULO

1970: It's Your Thing

LEE GARMES (1898-1978)

1939: Dreamy Lips; The Sky Is the Limit
1953: Outlaw Territory (w/ John Ireland)

TAY GARNETT (1895-1977)

1929: The Spieler; The Flying Fool
1930: Officer O'Brien; Her Man
1931: Bad Company
1932: Destination Unknown; One Way Passage; Okay America; Prestige
1933: S.O.S. Iceberg
1935: China Seas; Professional Soldier; She Couldn't Take It
1937: Love Is News; Slave Ship; Stand-In
1938: Joy of Living; Trade Winds
1939: Eternally Yours
1940: Seven Sinners; Slightly Honorable
1941: Cheers for Miss Bishop
1942: My Favorite Spy
1943: Bataan; The Cross of Lorraine
1944: Mrs. Parkington
1945: The Valley of Decision
1946: The Postman Always Rings Twice
1947: Wild Harvest
1949: A Connecticut Yankee in King Arthur's Court
1950: The Fireball
1951: Cause for Alarm; Soldiers Three
1952: One Minute to Zero
1953: Main Street to Broadway
1954: The Black Knight
1956: Seven Wonders of the World (w/ others)
1960: Night Fighters (Brit.)

1963: Cattle King
1970: The Delta Factor
1972: The Mad Trapper
1973: Timber Tramp
1976: Challenge to Be Free

OLIVER J. P. GARRETT

1942: Careful, Soft Shoulders

OTIS GARRETT (-1941)

1937: The Black Doll
1938: The Last Express; Personal Secretary; Danger on the Air; Lady in the Morgue
1939: The Witness Vanishes; Mystery of the White Room; Exile Express
1940: Margie (w/ P. G. Smith)
1941: Sandy Gets Her Man (w/ P. G. Smith)

GREG GARRISON

1961: Hey, Let's Twist!
1962: Two Tickets to Paris!

HARRY GARSON

1933: Beast of Borneo

LOUIS J. GASNIER (1882-1963) Fr.

1929: Darkened Rooms
1930: Slightly Scarlet (w/ E. H. Knopf); The Shadow of the Law (w/ M. Marcin); Mysterious Mr. Parkes (Fr.); The Virtuous Sin (w/ G. Cukor)
1931: The Lawyer's Secret (w/ M. Marcin); Silence (w/ M. Marcin)
1932: The Strange Case of Clara Deane (w/ M. Marcin); Forgotten Commandments (w/ W. Schoor)
1933: Gambling Ship (w/ M. Marcin)
1935: The Last Outpost (w/ C. Barton)
1937: The Gold Racket; Bank Alarm
1940: The Burning Question (Tell Your Children; Reefer Madness); Murder on the Yukon
1941: Stolen Paradise
1942: Fight On, Marines

LEON GAST

1977: The Grateful Dead (doc.)

ROBERTO GAVALDON (1909-) Mex.

(American films only)
1948: Adventures of Casanova
1955: The Littlest Outlaw

BRUCE GELLER

1973: Harry in Your Pocket

EDWARD GENOCK

1950: Cassino to Korea

GEORGE W. GEORGE

1957: The James Dean Story (doc. w/ R. Altman)

ANDREW GEORGIAS

1978: Bigtime

GERALD GERAGHTY

1948: Train to Alcatraz (w/ Philip Ford)

MAURICE GERAGHTY (1908-)

1951: The Sword of Monte Cristo

MARION GERING (1901-1977)

1931: I Take This Woman; Twenty-Four Hours
1932: Ladies of the Big House; Devil and the Deep; Madame Butterfly
1933: Pick Up; Jennie Gerhardt
1934: Good Dame; Thirty-Day Princess; Ready for Love
1935: Rumba
1936: Rose of the Rancho; Lady of Secrets

1937: Thunder in the City (Brit.)
1938: She Married an Artist
1950: Sarumba (Cuba)
1963: Violated Paradise (Jap.)

CLYDE GERONIMI

1951: Alice in Wonderland (w/ H. Luske, W. Jackson)

TED GERSHUNY

1974: Silent Night Bloody Night

HARRY GERSTAD

1960: 13 Fighting Men

NICOLAS GESSNER

1971: Someone Behind the Door
1976: The Little Girl Who Lives Down the Lane

CEDRIC GIBBONS (1893-1960)

1934: Tarzan and His Mate (w/ Jack Conway)

ALAN GIBSON

1972: Crescendo
1977: Checkered Flag or Crash

TOM GIBSON

1937: The Singing Buckaroo; Santa Fe Rides

PHILIP GILBERT

1971: Blood and Lace

DAVID GILER

1975: The Black Bird

JEFF GILLEN

1974: Deranged (w/ Alan Ormsby)

STUART GILMORE

1946: The Virginian
1951: Hot Lead
1952: The Half-Breed; Target; Captive Women

FRANK D. GILROY

1971: Desperate Characters
1977: From Noon 'Til Three
1978: Once in Paris

PETER GIMBEL

1971: Blue Water, White Death (doc.)

MILTON MOSES GINSBERG

1969: Coming Apart
1973: Werewolf of Washington

BERNARD GIRARD (c.1929-   )

1957: Ride Out for Revenge; The Green-Eyed Blonde
1958: The Party Crashers; As Young As We Are
1966: Dead Heat on a Merry-Go-Round
1969: The Mad Room
1972: The Happiness Cage
1975: Gone with the West

WILLIAM GIRDLER

1974: The Zebra Killer; Abby
1975: Sheba Baby; Asylum of Satan
1976: Grizzly
1977: Day of the Animals

ROBERT GIST

1966: An American Dream

WYNDHAM GITTENS

1938: Forbidden Valley

(Serials)

1938: Tim Tyler's Luck (w/ F. Beebe, C. Smith)

BENJAMIN GLAZER (1887-1958)

1929: Strange Cargo (w/ Arthur Gregor)
1948: Song of My Heart

JAMES GLEASON (1886-1959)

1935: Hot Tip (w/ R. McCarey)

FRANK GLENDON

1935: Circle of Death

JACK GLENN

1950: Cry Murder
1952: If Moscow Strikes (doc.)

BERT GLENNON (1893-1967)

1929: Syncopation
1930: Around the Corner; Girl of the Port; Paradise Island
1931: In Line of Duty
1932: South of Santa Fe

PETER GLENVILLE (1913-) Brit.

(American films only)
1958: Me and the Colonel
1961: Summer and Smoke
1967: The Comedians

JIM GLICKENHAUS

1979: The Astrologer

PETER GODFREY (1899-1970)

1939: The Lone Wolf Spy Hunt
1941: Unexpected Uncle
1942: Highways by Night
1944: Make Your Own Bed
1945: Hotel Berlin; Christmas in Connecticut
1946: One More Tomorrow (The Animal Kingdom)
1947: The Two Mrs. Carrolls; Cry Wolf; That Hagen Girl; Escape Me Never
1948: The Woman in White; The Decision of Christopher Blake
1949: The Girl from Jones Beach; One Last Fling
1950: Barricade; The Great Jewel Robbery; He's a Cockeyed Wonder
1952: One Big Affair
1956: Please Murder Me

JILL GODMILOW

1974: Antonia: A Portrait of a Woman (doc. w/ Judy Collins)

MENAHEM GOLAN (1929-) Isr.

(American films only)
1975: Lepke; Diamonds (Isr./U.S.)

WILLIS GOLDBECK (1900-)

1942: Dr. Gillespie's New Assistant
1943: Dr. Gillespie's Criminal Case
1944: Rationing; Three Men in White; Between Two Women
1945: She Went to the Races
1946: Love Laughs at Andy Hardy
1947: Dark Delusion
1949: Johnny Holiday
1951: Ten Tall Men

JACK GOLDBERG

1944: We've Come a Long Way

MARTIN GOLDMAN

1972: The Legend of Nigger Charley

JAMES GOLDSTONE (1931-)

1968: Jigsaw

1969: A Man Called Gannon; Winning
1971: Brother John; Red Sky at Morning; The Gang That Couldn't Shoot Straight
1972: They Only Kill Their Masters

PHIL GOLDSTONE

1933: The Sin of Nora Moran
1938: Marriage Forbidden

RICHARD GOLDSTONE

1958: South Seas Adventure (w/ others)
1962: No Man Is an Island (w/ John Monks, Jr.)

SAMUEL GOLDWYN, JR. (1926-   )

1964: The Young Lovers

SERVANDO GONZALEZ

1965: The Fool Killer

SAUL A. GOODKIND

(Serials)

1939: Buck Rogers (w/ F. Beebe); The Phantom Creeps (w/ F. Beebe); The Oregon Trail (w/ F. Beebe)

EDWARD GOODMAN

1931: Women Love Once

ROBERT L. GOODWIN

1971: Black Chariot

LESLIE GOODWINS (1889-1969)

1936: With Love and Kisses
1937: Anything for a Thrill; Headline Crasher; Young Dynamite
1938: Crime Ring; Fugitives for a Night; Mr. Doodle Kicks Off; Tarnished Angel; Glamor Boy

1939: The Girl From Mexico; Mexican Spitfire; The Day the Bookies Wept; Almost a Gentleman; Sued for Libel
1940: Men Against the Sky; Millionaire Playboy; Pop Always Pays; Let's Make Music; Mexican Spitfire Out West
1941: Parachute Battalion; They Met in Argentina (w/ Jack Hively); Mexican Spitfire's Baby
1942: Mexican Spitfire at Sea; Mexican Spitfire Sees a Ghost; Mexican Spitfire's Elephant
1943: Silver Skates; Ladies' Day; Gals, Inc.; Mexican Spitfire's Blessed Event; The Adventures of a Rookie; Rookies in Burma
1944: Casanova in Burlesque; The Mummy's Curse; Murder in the Blue Room; The Singing Sheriff; Goin' to Town; Hi, Beautiful
1945: I'll Tell the World; Radio Stars on Parade; What a Blonde; An Angel Comes to Brooklyn
1946: Genius at Work; Riverboat Rhythm; Vacation in Reno
1947: The Lone Wolf in London; Dragnet
1952: Gold Fever
1954: Fireman Save My Child
1955: Fresh From Paris; Paris Follies of 1956
1967: Tammy and the Millionaire (w/ Sidney Miller, Ezra Stone)

BERT I. GORDON (1922-)

1955: King Dinosaur
1957: Beginning of the End; The Cyclops; The Amazing Colossal Man
1958: Attack of the Puppet People; War of the Colossal Beast; The Spider
1960: The Boy and the Pirates; Tormented
1962: The Magic Sword
1965: Village of the Giants
1966: Picture Mommy Dead
1970: How to Succeed with Sex
1972: Necromancy
1973: The Mad Bomber; The Police Connection
1976: Food of the Gods
1977: Empire of the Ants

GERALD GORDON

1973: So Long, Blue Boy

MICHAEL GORDON (1909-)

1942: Boston Blackie Goes Hollywood; Underground Agent
1943: One Dangerous Night; Crime Doctor
1947: The Web
1948: Another Part of the Forest; An Act of Murder

1949: The Lady Gambles; Woman in Hiding
1950: Cyrano de Bergerac
1951: I Can Get It for You Wholesale; The Secret of Convict Lake
1953: Wherever She Goes (Australia)
1959: Pillow Talk
1960: Portrait in Black
1962: Boys' Night Out
1963: For Love or Money; Move Over Darling
1965: A Very Special Favor
1966: Texas Across the River
1968: The Impossible Years
1970: How Do I Love Thee?

ROBERT GORDON

1947: Black Eagle; Blind Spot; Sport of Kings
1953: The Joe Louis Story
1955: It Came from Beneath the Sea
1958: Damn Citizen; The Rawhide Trail
1963: Black Zoo
1968: Tarzan and the Jungle Boy
1972: The Gatling Gun

BERRY GORDY

1975: Mahogany

LARRY GOTTHEIM

1974: Horizons (doc.)

CHARLES S. GOULD

1955: Jungle Moon Men

(Serial)

1953: The Great Adventures of Capt. Kidd (w/ D. Abrahams)

DAVE GOULD

1942: Rhythm Parade

STANLEY GOULDER

1975: Naked Evil

ALFRED GOULDING

1929: All at Sea
1940: A Chump at Oxford

EDMUND GOULDING (1891-1959)

1929: The Trespasser
1930: The Devil's Holiday; Paramount on Parade (w/ others)
1931: The Night Angel; Reaching for the Moon
1932: Blondie of the Follies; Grand Hotel
1934: Riptide
1935: The Flame Within
1937: That Certain Woman
1938: The Dawn Patrol; White Banners
1939: Dark Victory; The Old Maid; We Are Not Alone
1940: 'Til We Meet Again
1941: The Great Lie
1943: Claudie; The Constant Nymph; Forever and a Day (w/ others)
1946: The Razor's Edge; Of Human Bondage
1947: Nightmare Alley
1949: Everybody Does It
1950: Mister 880
1952: We're Not Married
1953: Down Among the Sheltering Palms
1956: Teen-age Rebel
1958: Mardi Gras

TOM GRAEFF (1929-)

1959: Teenagers from Outer Space

JO GRAHAM

1942: Always in My Heart; You Can't Escape Forever
1943: The Good Fellows

WALTER GRAHAM

1929: Divorce Made Easy

WILLIAM A. GRAHAM

1967: Waterhole #3
1969: Change of Habit
1971: Honky
1972: Face to the Wind (alt. Count Your Bullets)

1974: Together Brothers; Where the Lilies Bloom
1976: Sounder, Part 2

RICHARD GRAND

1976: The Commitment
1979: Fyre

JAMES EDWARD GRANT (1902-1966)

1947: Angel and the Badman
1954: Ring of Fear

ALEX GRASSHOFF

1967: Young Americans

WALTER E. GRAUMAN (1922-)

1957: The Disembodied
1964: Lady in a Cage; 633 Squadron
1965: A Rage to Live
1966: I Deal in Danger
1970: The Last Escape (Germ.)

GARY GRAVER

1963: The Great Dream
1970: The Hard Road; Erika's Hot Summer; Sandra--The Making of a Woman
1973: There Was a Little Girl

WILLIAM GREAVES

1974: The Fighters
1975: Ali the Man: Ali the Fighter (doc. w/ Rick Baxter)

ALFRED E. GREEN (1889-1960)

1929: Making the Grade; Disraeli
1930: The Green Goddess; The Man from Blankley's; Old English; Sweet Kitty Bellaire
1931: Smart Money; Men of the Sky; The Road to Singapore
1932: Union Depot; It's Tough to Be Famous; The Rich Are Always with Us; The Dark Horse; Silver Dollar

1933: Parachute Jumper; The Narrow Corner; Baby Face; I Loved a Woman
1934: As the Earth Turns, Dark Hazard; The Merry Frinks; Housewife; Side Streets; A Lost Lady; Gentlemen Are Born
1935: Sweet Music; The Girl from 10th Avenue; Here's to Romance; The Goose and the Gander; Dangerous
1936: Colleen; The Golden Arrow; Two in a Crowd; They Met in a Taxi; More Than a Secretary
1937: Let's Get Married; The League of Frightened Men; Mr. Dodd Takes the Air; Thoroughbreds Don't Cry
1938: Ride a Crooked Mile; Duke of West Point
1939: The King of the Turf; The Gracie Allen Murder Case; 20,000 Men a Year
1940: Shooting High; South of Pago Pago; Flowing Gold; East of the River
1941: Adventure in Washington; Badlands of Dakota
1942: The Mayor of Forty-fourth Street; Meet the Stewarts
1943: Appointment in Berlin; There's Something About a Soldier
1944: Mr. Winkle Goes to War; Strange Affair
1945: A Thousand and One Nights
1946: Tars and Spars; The Jolson Story
1947: The Fabulous Dorseys; Copacabana
1948: Four Faces West; The Girl from Manhattan
1949: Cover-up
1950: Sierra; The Jackie Robinson Story
1951: Two Gals and a Guy
1952: Invasion
1953: Paris Model; The Eddie Cantor Story
1954: Top Banana

GUY GREEN (1913-) Brit.

(American films only)
1962: Light in the Piazza
1963: Diamond Head
1965: A Patch of Blue
1970: Walk in the Spring Rain
1975: Once Is Not Enough

WALON GREEN

1967: Spree (doc. w/ M. Leisen)
1971: The Hellstrom Chronicle (doc.)
1979: The Secret Life of Plants

DAVID GREENE (1924-) Brit.

(American films only)
1970: The People Next Door

1973: Godspell
1978: Gray Lady Down

DAVID ALLEN GREENE

1968: Come Back Baby

FELIX GREENE

1971: Cuba Va (doc.)

HERBERT GREENE

1959: The Cosmic Man

WILLIAM GREFE

1964: Racing Fever
1969: Hooked Generation
1973: Stanley
1975: Impulse
1976: The Jaws of Death

ARTHUR GREGOR

1929: Strange Cargo (w/ B. Glazer)
1933: What Price Decency?

ABNER J. GRESHLER

1953: Yesterday and Today (doc.)

HARRY WAGSTAFF GRIBBLE

1932: Madame Racketeer (w/ A. Hall)

TOM (THOMAS S.) GRIES (1922-)

1955: Hell's Horizon
1958: Girl in the Woods
1968: Will Penny
1969: 100 Rifles; Number One
1970: The Hawaiians; Fools
1972: Journey Through Rosebud
1973: Lady Ice

1975: Dynamite Man; Breakout (Span./Fr./U.S.)
1976: Breakheart Pass
1977: The Greatest

CHARLES B. GRIFFITH

1959: Forbidden Island
1976: Eat My Dust!

DAVID WARK GRIFFITH (1875-1948)

1929: Lady of the Pavements
1930: Abraham Lincoln
1931: The Struggle
1940: One Million B.C. (unc. w/ Hal Roach, Hal Roach, Jr.)

EDWARD H. GRIFFITH (1894-   )

1929: Paris Bound; The Shady Lady; Rich People
1930: Holiday
1931: Rebound
1932: The Animal Kingdom; Lady with a Past
1933: Another Language
1935: Biography of a Bachelor Girl; No More Ladies
1936: Ladies in Love; Next Time We Love
1937: I'll Take Romance; Cafe Metropole
1939: Honeymoon in Bali; Cafe Society
1940: Safari
1941: Bahama Passage; One Night in Lisbon; Virginia
1943: The Sky's the Limit; Young and Willing
1946: Perilous Holiday

HUGO GRIMALDI

1959: Gigantis, the Fire Monster (Jap.)
1965: The Human Duplicators; Mutiny in Outer Space

NICK (HARRY A.) GRINDE (1893-1979)

1929: Morgan's Last Raid; The Desert Rider
1930: The Bishop Murder Case (w/ David Burton); Good News (w/ Edgar J. McGregor); Remote Control
1931: This Modern Age
1932: Shopworn; Vanity Street
1935: Stone of Silver Creek; Border Brigands; Ladies Crave Excitement
1936: Jailbreak; Public Enemy's Wife
1937: Fugitive in the Sky; The Captain's Kid; White Bondage; Public Wedding; Love Is on the Air; Exiled to Shanghai

1938: Down in Arkansas
1939: Federal Man-Hunt; King of Chinatown; Sudden Money; Million Dollar Legs; The Man They Could Not Hang; A Woman Is the Judge
1940: Scandal Sheet; Convicted Woman; The Man with Nine Lives; Men Without Souls; Girls of the Road; Before I Hang; Friendly Neighbors
1942: The Girl from Alaska
1943: Hitler--Dead or Alive
1945: Road to Alcatraz

WALLACE A. GRISSELL (1904-)

1944: Marshal of Reno; Vigilantes of Dodge City
1945: Wanderer of the Wasteland (w/ E. Killy); Corpus Christi Bandits
1947: Wild Horse Mesa
1948: Western Heritage
1952: A Yank in Indo-China

(Serials)

1944: Haunted Harbor (w/ S. Bennet); The Tiger Woman; Zorro's Black Whip (both w/ S. Bennet)
1945: Federal Operator 99; Manhunt on Mystery Island (both w/ S. Bennet, Y. Canutt); Who's Guilty? (w/ H. Bretherton)
1951: Captain Video; King of the Congo (both w/ S. Bennet)

JOHN GRISSMER

1976: Scalpel

FERDE GROFE, JR.

1968: Warkill
1972: The Proud and the Damned

ULU GROSBARD (1929-)

1968: The Subject Was Roses
1971: Who Is Harry Kellerman and Why Is He Saying Those Terrible Things About Me?
1978: Straight Time

JERRY GROSS

1968: Teenage Mother

JAMES WILLIAM GUERCIO

1973: Electra Glide in Blue

CHARLES GUGGENHEIM

1959: The Great St. Louis Bank Robbery (w/ J. Stix)

KING GUIDICE

1937: Timberesque

PAUL GUILFOYLE

1960: Tess of the Storm Country

JOHN GUILLERMIN (1925-) Brit.

(American films only)
1966: The Blue Max
1968: P. J.
1969: The Bridge at Remagen; House of Cards
1970: El Condor
1972: Skyjacked
1973: Shaft in Africa
1974: The Towering Inferno (w/ Irwin Allen)
1976: King Kong
1978: Death on the Nile

FRED GUIOL

1934: What's Your Racket?
1935: The Rainmakers
1936: Silly Billies; Mummy's Boys
1941: Tanks a Million; Miss Polly
1942: Hay Foot
1946: Here Comes Trouble
1951: As Your Were

BILL GUNN

1973: Ganja and Hess

CHARLES HAAS (1913-   )

1956: Star in the Dust; Screaming Eagles; Showdown at Abilene
1958: Summer Love; Wild Heritage
1959: The Beat Generation; The Big Operator; Girls Town
1960: Platinum High School

HUGO HAAS (1901-1968) Czech

1951: Pickup; The Girl on the Bridge
1952: Strange Fascination
1953: One Girl's Confession; Thy Neighbor's Wife
1954: Bait; The Other Woman
1955: Hold Back Tomorrow
1956: Edge of Hell
1957: Lizzie; Hit and Run
1959: Night of the Quarter Moon; Born to Be Loved

GEORGE HADDEN

1934: Charlie Chan's Courage

HORST HAECHLER

1960: As the Sea Rages

ROSS HAGEN

1979: The Glove

MACK HAGGARD

1977: The First Nudie Musical (w/ Bruce K. Mamel)

LARRY HAGMAN (1939-   )

1972: Beware! the Blob

STUART HAGMANN

1970: The Strawberry Statement
1971: Believe in Me

FRED HAINES

1974: Steppenwolf

REX HALE

1936: Racing Blood

WILLIAM HALE

1967: Gunfight in Abilene
1968: Journey to Shiloh

EARL HALEY

1933: King of the Wild Horses
1939: The Gentleman from Arizona

JACK HALEY, JR. (1933-)

1970: Norwood
1971: The Love Machine
1974: That's Entertainment

H. B. HALICKI

1974: Gone in 60 Seconds

ALEXANDER HALL (1894-1968)

1932: Sinners in the Sun; Madame Racketeer (w/ Harry Wagstaff Gribble)
1933: The Girl in 419 (w/ George Somnes); Midnight Club (w/ G. Somnes); Torch Singer (w/ G. Somnes)
1934: Miss Fane's Baby Is Stolen; Little Miss Marker; The Pursuit of Happiness; Limehouse Blues
1935: Goin' to Town; Annapolis Farewell
1936: Give Us This Night; Yours for the Asking
1937: Exclusive
1938: There's Always a Woman; I Am the Law; There's That Woman Again
1939: The Lady's from Kentucky; Good Girls Go to Paris; The Amazing Mr. Williams
1940: The Doctor Takes a Wife; He Stayed for Breakfast; This Thing Called Love; Here Comes Mr. Jordan; Bedtime Story
1942: They All Kissed the Bride; My Sister Eileen
1943: The Heavenly Body
1944: Once Upon a Time
1945: She Wouldn't Say Yes
1947: Down to Earth
1949: The Great Lover

1950: Love That Brute; Louisa
1951: Up Front
1952: Because You're Mine
1953: Let's Do It Again
1956: Forever Darling

JON HALL (1913-)

1963: Monster from the Surf

DANIEL HALLER (1926-)

1965: Die, Monster, Die!
1967: Devil's Angels
1968: The Wild Racers
1970: Paddy (Ire.); The Dunwich Horror; Pieces of Dreams
1979: Buck Rogers in the 25th Century

RICHARD HALLIBURTON

1933: India Speaks (doc.)

VICTOR HUGO HALPERIN (1895-)

1930: Party Girl
1931: Ex-Flame
1932: White Zombie
1933: Supernatural
1936: I Conquer the Sea
1937: Nation Aflame
1939: Torture Ship
1940: Buried Alive
1942: Girls Town

WILLIAM HAMILTON (-1942)

1935: Freckles (w/ Edward Killy); Seven Keys to Baldpate (w/ E. Killy)
1936: Murder on the Bridle Path (w/ E. Killy); Bunker Bean (w/ E. Killy)
1941: Call Out the Marines (w/ Frank Ryan)

ALEX HAMMID

1951: Of Men and Music (w/ Irving Reis)

JOHN HANCOCK (1939-)

1971: Let's Scare Jessica to Death
1973: Bang the Drum Slowly
1976: Baby Blue Marine
1979: California Dreaming

KEN HANDLER

1974: A Place Without Parents (orig. Pigeons)
1975: Truckin'

JIM HANDLEY

1941: The Reluctant Dragon (anim. w/ others)

WILLIAM HANNA (1910-)

1964: Hey There It's Yogi Bear (anim. w/ Joseph Barbera)
1966: The Man Called Flintstone (anim. w/ Joseph Barbera)

CURTIS HANSON

1973: The Arousers

JOSEPH C. HANWRIGHT

1978: Uncle Joe Shannon

DEAN HARGROVE

1975: The Manchu Eagle Murder Caper Mystery

RICHARD HARLAN

1938: Radio Troubador; Bachelor Father; Papa Soltero
1940: Mercy Plane

CURTIS HARRINGTON (1928-)

1963: Night Tide
1966: Queen of Blood
1967: Games
1971: What's the Matter with Helen?; Who Slew Auntie Roo?

1973: The Killing Kind

DENNY HARRIS

1979: Silent Scream

JAMES B. HARRIS (1928-)

1965: The Bedford Incident
1973: Some Call It Loving

HAL HARRISON, JR.

1976: Pony Express Rider

PAUL HARRISON

1974: The House of Seven Corpses

HARVEY HART (1928-)

1965: Bus Riley's Back in Town; Dark Intruder
1967: Sullivan's Empire (w/ Thomas Carr)
1968: The Sweet Ride
1971: Fortune and Men's Eyes
1973: The Pyx (Can.)
1976: Shoot (Can.)
1977: Goldenrod (Can.)
1979: The Mad Trapper

WALTER HART

1950: The Goldbergs (ret. Molly)

ROBERT HARTFORD-DAVIS

1972: Black Gunn
1974: The Take

DON HARTMAN (1900-1958)

1947: It Had to Be You (w/ Rudolph Mate)
1948: Every Girl Should Be Married
1949: Holiday Affair
1951: It's a Big Country (w/ others); Mr. Imperium

EMILE HARVARD

1975: Fugitive Killer

ANTHONY HARVEY (1931-   ) Brit.

(American films only)
1967: Dutchman
1971: They Might Be Giants
1979: Players; Eagle's Wing

LAURENCE HARVEY (1928-1973) Brit.

1963: The Ceremony
1974: Welcome to Arrow Beach

BYRON HASKIN (1899-   )

1947: I Walk Alone
1948: Man-Eater of Kumaon
1949: Too Late for Tears
1950: Treasure Island (Brit.)
1951: Tarzan's Peril; Warpath; Silver City
1952: The Denver and Rio Grande
1953: The War of the Worlds; His Majesty O'Keefe
1954: The Naked Jungle
1955: Conquest of Space; Long John Silver (Australia)
1956: The First Texan; The Boss
1958: From the Earth to the Moon
1959: The Little Savage
1960: Jet over the Atlantic; September Storm
1961: Armored Command
1963: Captain Sinbad
1964: Robinson Crusoe on Mars
1968: The Power

HENRY HATHAWAY (1898-   )

1932: Wild Horse Mesa
1933: Under the Tonto Rim; Heritage of the Desert; Man of the Forest; Sunset Pass; The Thundering Herd; To the Last Man
1934: Now and Forever; Come on Marines; The Witching Hour; The Last Round-Up
1935: The Lives of a Bengal Lancer; Peter Ibbetson
1936: Go West, Young Man; Trail of the Lonesome Pine; I Loved a Soldier (unfinished)
1937: Souls at Sea
1938: Spawn of the North

1939: The Real Glory
1940: Brigham Young--Frontiersman; Johnny Apollo
1941: Shepherd of the Hills; Sundown
1942: China Girl; Ten Gentlemen from West Point
1944: Home in Indiana; Wing and a Prayer
1945: The House on 92nd Street; Nob Hill
1946: The Dark Corner; 13 Rue Madeleine
1947: Kiss of Death
1948: Call Northside 777 (ret. Calling Northside 777)
1949: Down to the Sea in Ships
1950: The Black Rose
1951: The Desert Fox; 14 Hours; Rawhide; U.S.S. Teakettle
1952: Diplomatic Courier; O. Henry's Full House (w/ others)
1953: Niagara; White Witch Doctor
1954: Garden of Evil; Prince Valiant
1955: The Racers
1956: The Bottom of the Bottle; 23 Paces to Baker Street
1957: Legend of the Lost
1958: From Hell to Texas
1959: Woman Possessed
1960: North to Alaska; Seven Thieves
1963: How the West Was Won (w/ others)
1964: Circus World
1965: The Sons of Katie Elder
1966: Nevada Smith
1967: The Last Safari (Brit.)
1968: Five Card Stud
1969: True Grit
1971: Raid on Rommel; Shoot-Out
1974: Hangup

STEVE HAWKES

1975: Steve, Samson and Delilah

HOWARD HAWKS (1896-1977)

1930: The Dawn Patrol
1931: The Criminal Code
1932: The Crowd Roars; Tiger Sharks; Scarface
1933: Today We Live; The Prizefighter and the Lady (unc. w/ W. S. Van Dyke)
1934: Twentieth Century; Viva Villa (unc. w/ Jack Conway)
1935: Barbary Coast; Ceiling Zero
1936: Come and Get It (w/ W. Wyler); The Road to Glory
1938: Bringing Up Baby
1939: Only Angels Have Wings
1940: His Girl Friday
1941: Ball of Fire; Sergeant York
1943: Air Force; The Outlaw (unc. w/ Howard Hughes)
1944: To Have and Have Not

1946: The Big Sleep
1948: Red River; A Song Is Born
1949: I Was a Male War Bride
1951: The Thing (unc. w/ C. Nyby)
1952: The Big Sky; Monkey Business; O. Henry's Full House (w/ others)
1953: Gentlemen Prefer Blondes
1955: Land of the Pharaohs
1959: Rio Bravo
1962: Hatari!
1964: Man's Favorite Sport?
1965: Red Line 7000
1966: El Dorado
1970: Rio Lobo

KENNETH HAWKS (-1930)

1929: Big Time; Masked Emotions (w/ David Butler)
1930: Such Men Are Dangerous

JEFFREY HAYDEN

1957: The Vintage

RICHARD HAYDN (1905-)

1948: Miss Tatlock's Millions
1949: Dear Wife
1950: Mr. Music

JOHN HAYES

1974: Garden of the Dead; Grave of the Vampire
1977: End of the World

MANNING HAYNES

1931: Should a Doctor Tell?

PAUL F. HEARD

1958: Hong Kong Affair

ARCH HEATH

1929: Modern Love

BEN HECHT (1894-1964)

1934: Crime Without Passion (w/ Charles MacArthur)
1935: The Scoundrel (w/ C. MacArthur)
1936: Soak the Rich (w/ C. MacArthur); Once in a Blue Moon (w/ C. MacArthur)
1940: Angels over Broadway (w/ Lee Garmes)
1946: Specter of the Rose
1952: Actors and Sin

VICTOR HEERMAN (1892-)

1930: Personality; Paramount on Parade (w/ others); Animal Crackers; Sea Legs

RICHARD HEFFRON

1972: Fillmore (doc.)
1974: Newman's Law
1976: Futureworld; Trackdown
1977: Outlaw Blues

MORTON HEILIG

1974: Once

RUSSELL RAY HEINZ

1935: Blazing Guns; Border Vengeance
1936: Just My Luck

STUART HEISLER (1894-1980)

1936: Straight From the Shoulder
1937: The Hurricane (unc. w/ J. Ford)
1940: The Biscuit Eater
1941: Among the Living; The Monster and the Girl
1942: The Glass Key; The Remarkable Andrew
1945: Along Came Jones
1946: Blue Skies
1947: Smash-Up, the Story of a Woman
1949: Tokyo Joe; Tulsa
1950: Chain Lightning; Dallas; Storm Warning; Vendetta (unc. w/ others)
1951: Journey into Light
1952: Island of Desire
1953: The Star
1954: Beachhead; This Is My Love

1955: I Died a Thousand Times
1956: The Burning Hills; The Lone Ranger
1962: Hitler

JEROME HELLMAN

1979: Promises in the Dark

MONTE HELLMAN (1932-)

1959: The Beast from Haunted Cave
1965: Flight to Fury (Philippines); Back Door to Hell (Philippines)
1971: The Shooting; Ride in the Whirlwind; Two-Lane Blacktop
1974: Shatter (Hong Kong); Cockfighter
1978: China 9 Liberty 37 (Ital.)

OLIVER HELLMAN

1977: Tentacles

GUNNAR HELLSTROM

1968: The Name of the Game Is Kill

PAUL HELMICK

1960: Thunder in Carolina

DAVID HELPERN, JR.

1974: I'm a Stranger Here Myself (doc.)
1976: Hollywood on Trial (doc.)
1979: Something Short of Paradise

JOSEPH E. HENABERY (1888-1976)

1929: Red Hot Speed; Clear the Decks; Light Fingers
1930: The Love Trader
1943: The Leather Burners

RALPH HENABERY

1935: Speed Devils

HOBART HENLEY

1928: A Certain Young Man
1929: The Lady Lies
1930: The Big Pond; Roadhouse Nights; Mothers Cry; Free Love
1931: Captain Applejack; Bad Sisters; Expensive Women
1932: Night World
1934: Unknown Blonde

PAUL HENREID (1907-)

1952: For Men Only
1956: A Woman's Devotion
1958: Girls on the Loose; Live Fast, Die Young
1964: Dead Ringer
1965: Blues for Lovers (Brit.; orig. Ballad in Blue)

BUCK HENRY (1930-)

1978: Heaven Can Wait (w/ Warren Beatty)

F. HUGH HERBERT (1897-1958)

1930: He Knew Women
1948: Scudda Hoo! Scudda Hay!
1953: The Girls of Pleasure Island (w/ A. Ganzer)

MARTIN HERBERT

1977: Strange Shadows in an Empty Room

ALBERT HERMAN

1931: Sporting Chance
1932: Exposed
1933: The Big Chance
1935: Western Frontier; Danger Ahead; Gun Play; Hot off the Press; Trail's End; What Price Crime? Twisted Rails; Cowboy and the Bandit; Big Boy Rides Again
1936: Blazing Justice; Outlaws of the Range; Bars of Hate
1937: Renfrew of the Royal Mounted; Valley of Terror
1938: Starlight Over Texas; Where the Buffalo Roam; Rollin' Plains; On the Great White Trail; Utah Trail
1939: Down the Wyoming Trail; Roll Wagons Roll; Song of the Buckaroo; Sundown on the Prairie; Rollin' Westward; Man from Texas
1940: Arizona Frontier; The Golden Trail; Pals of the Silver Sage;

Rainbow over the Range; Rhythm of the Rio Grande; Take Me Back to Oklahoma
1941: Rolling Home to Texas; The Pioneers; Gentleman from Dixie
1942: Nazi Spy Ring; A Yank in Libya
1943: The Rangers Take Over; Miss V from Moscow
1944: Delinquent Daughters; Shake Hands with Murder
1945: The Missing Corpse; The Phantom of 42nd Street; Rogue's Gallery

(Serials)

1933: The Whispering Shadow (w/ C. Clark)
1936: The Clutching Hand

NORMAN HERMAN

1959: Tokyo After Dark

DENIS HEROUX

1975: Jacques Brel Is Alive and Well and Living in Paris

HERRICK F. HERRICK

1934: Obeah (doc.)
1935: Black Hell

BRUCE HERSCHENSOHN

1966: John F. Kennedy: Years of Lightning, Day of Drums

GORDON HESSLER

1965: The Woman Who Wouldn't Die
1969: The Oblong Box (Brit.); The Last Shot You Hear (Brit.)
1970: The Cry of the Banshee (Brit.); Scream and Scream Again (Brit.)
1971: Murders in the Rue Morgue
1974: The Golden Voyage of Sinbad (Brit.)

DAVID L. HEWITT

1967: Journey to the Center of Time; Return from the Past

DOUGLAS HEYES

1964: Kitten with a Whip
1966: Beau Geste

JESSIE HIBBS (1906-)

1953: The All American
1954: Ride Clear of Diablo; Black Horse Canyon; Rails into Laramie; The Yellow Mountain
1955: To Hell and Back; The Spoilers
1956: World in My Corner; Walk the Proud Land
1957: Joe Butterfly
1958: Ride a Crooked Trail

DOUGLAS HICKOX

1959: The Giant Behemoth (Brit. w/ E. Lourie)
1966: Disk-O-Tek Holiday
1975: Brannigan
1976: Sky Riders

HOWARD HIGGIN (1893-1937)

1929: Sal of Singapore; The Leatherneck; High Voltage
1930: The Racketeer
1931: The Painted Desert
1932: The Last Man; The Final Edition; Hell's House
1933: Carnival Lady; Marriage on Approval
1934: The Line-Up
1937: Battle of Greed

COLIN HIGGINS

1978: Foul Play

NAT HIKEN

1969: The Love God?

GEORGE ROY HILL (1922-)

1962: Period of Adjustment
1963: Toys in the Attic
1964: The World of Henry Orient
1966: Hawaii
1967: Thoroughly Modern Millie
1969: Butch Cassidy and the Sundance Kid
1972: Slaughterhouse Five
1973: The Sting
1975: The Great Waldo Pepper
1977: Slap-Shot
1979: A Little Romance

GEORGE W. HILL (1895-1934)

1929: The Flying Fleet
1930: The Big House; Min and Bill
1931: The Secret Six; Hell Divers
1933: Clear All Wires

JACK HILL

1966: Track of the Vampire
1969: Pit Stop
1971: The Big Doll House
1972: The Big Bird Cage
1973: Coffy
1974: Foxy Brown; The Swinging Cheerleaders
1975: Switchblade Sisters

JAMES A. HILL (1919-)

1962: Trial and Error
1965: Seaside Swingers
1966: The Corrupt Ones; Born Free; A Study in Terror
1970: Captain Nemo and the Underwater City; Man from O.R.G.Y.

JEROME HILL

1957: Albert Schweitzer (doc.)
1960: The Sand Castle
1964: Open the Door and See All the People

ROBERT F. HILL (1886-)

1929: Melody Lane
1931: Sundown Trail
1932: Love Bound; Come On Danger
1933: Tarzan, the Fearless (adapt. from Serial); Cheyenne Kid
1934: A Demon for Trouble; Inside Information; Outlaw's Highway; Frontier Days; Cowboy Holiday
1935: Cyclone Rider; Texas Rambler; Vanishing Riders; Danger Trails
1936: The Idaho Kid; Kelly of the Secret Service; Men of the Plains; Prison Shadows; Put on the Spot; Rio Grande Romance; The Rogue's Tavern; Too Much Beef; West of Nevada; Face in the Fog; Taming the Wild; Rip Roaring Buckaroo; Law and Lead; Phantom of the Range
1937: Two Minutes to Play; Million Dollar Racket; Cheyenne Rides Again; Feud on the Trail; Mystery Range; The Roaming Cowboy
1938: Whirlwind Horseman; Flying Fists; Man's Country; The Painted Trail; Silks and Saddles

1939: Overland Mail; Wild Horse Canyon; Drifting Westward
1940: East Side Kids
1941: Wanderers of the West

(Serials)

1931: Spell of the Circus; Heroes of the Flames
1933: Tarzan the Fearless
1936: Shadow of Chinatown
1937: Blake of Scotland Yard
1938: Flash Gordon's Trip to Mars (w/ F. Beebe)

WALTER HILL (1942-)

1975: Hard Times
1978: The Driver
1979: The Warriors

ARTHUR HILLER (1923-)

1957: The Careless Years
1963: Miracle of the White Stallions; The Wheeler Dealers
1964: The Americanization of Emily
1966: Promise Her Anything (Brit.); Penelope
1967: Tobruk; The Tiger Makes Out
1969: Popi
1970: The Out-of-Towners; Love Story
1971: The Hospital; Plaza Suite
1972: Man of La Mancha (Ital.)
1974: The Crazy World of Julius Vrooder
1975: The Man in the Glass Booth
1976: Silver Streak; W. C. Fields and Me
1979: The In-Laws; Nightwing

RICHARD L. HILLIARD

1964: Psychomania

LAMBERT HILLYER (1889-)

1930: Beau Bandit
1932: The Deadline; One-Man Law; The Fighting Fool; South of the Rio Grande; White Eagle; Hello, Trouble; The Sundown Rider
1933: The Forbidden Trail; Dangerous Crossroads; The California Trail; Unknown Valley; Police Car 17; Before Midnight; Master of Men
1934: The Fighting Code; Once to Every Woman; One Is Guilty; The Guilty; The Man Trailer; The Defense Rests; Most Precious Thing in Life; Against the Law; Men of the Night

1935: Behind the Evidence; In Spite of Danger; Men of the Hour; Awakening of Jim Burke; Guard That Girl!; Superspeed
1936: The Invisible Ray; Dangerous Waters; Dracula's Daughter
1937: Speed to Spare; Girls Can Play; Women in Prison
1938: My Old Kentucky Home; All-American Sweetheart; Extortion
1939: Convict's Code; Should a Girl Marry?; Girl from Rio
1940: The Durango Kid
1941: The Pinto Kid; North from the Lone Star; The Wildcat of Tucson; The Return of Daniel Boone; Beyond Sacramento; Hands Across the Rockies; The Son of Davy Crockett; The Medico of Painted Springs; King of Dodge City; Prairie Stranger; Thunder over the Prairie; Roaring Frontiers; The Royal Mounted Patrol
1942: North of the Rockies; The Devil's Trail; Prairie Gunsmoke; Vengeance of the West
1943: Fighting Frontier; Six-Gun Gospel; The Stranger from Pecos; The Texas Kid
1944: Smart Guy; Partners of the Trail; Law Men; West of the Rio Grande; Land of the Outlaws; Ghost Guns
1945: Beyond the Pecos; Flame of the West; Stranger from Santa Fe; South of the Rio Grande; The Lost Trail; Frontier Feud
1946: Border Bandits; Under Arizona Skies; The Gentleman from Texas; Trigger Fingers; Shadows on the Range; Silver Range
1947: Raiders of the South; Valley of Fear; Trailing Danger; Land of the Lawless; The Law Comes to Gunsight; The Hat Box Mystery; The Case of the Baby Sitter; Flashing Guns; Prairie Express; Gun Talk
1948: Song of the Drifter; Overland Trails; Oklahoma Blues; Crossed Trails; Partners of the Sunset; Frontier Agent; Range Renegades; The Fighting Ranger; The Sheriff of Medicine Bow; Outlaw Brand
1949: Gun Runner; Gun Law Justice; Trails End; Haunted Trails; Riders of the Dusk; Range Land

(Serials)

1943: Batman

ARTHUR DAVID HILTON

1950: The Return of Jesse James
1954: Cat-Women of the Moon

ROBERT HINKLE

1961: Ole Rex

ALFRED HITCHCOCK (1899-1979)

(British films)
1929: Blackmail

1930: Murder; Juno and the Paycock
1931: The Skin Game
1932: East of Shanghai; Number Seventeen; Rich and Strange
1934: The Man Who Knew Too Much
1935: The 39 Steps; Strauss's Great Waltz
1936: Sabotage; The Secret Agent
1937: Young and Innocent
1938: The Lady Vanishes
1939: Jamaica Inn
(American films)
1940: Rebecca; Foreign Correspondent
1941: Mr. and Mrs. Smith; Suspicion
1942: Saboteur
1943: Shadow of a Doubt
1944: Lifeboat
1945: Spellbound
1946: Notorious
1947: The Paradine Case
1948: Rope
1949: Under Capricorn
1950: Stage Fright
1951: Strangers on a Train
1953: I Confess
1954: Dial M for Murder; Rear Window
1955: To Catch a Thief; The Trouble with Harry
1956: The Man Who Knew Too Much; The Wrong Man
1958: Vertigo
1959: North by Northwest
1960: Psycho
1963: The Birds
1964: Marnie
1966: Torn Curtain
1969: Topaz
1972: Frenzy (Brit.)
1976: Family Plot

CARL K. HITTLEMAN

1956: Kentucky Rifle
1957: Gun Battle at Monterey; The Buckskin Lady

JACK HIVELY

1939: They Made Her a Spy; Panama Lady; The Spellbinder; Three Sons; Two Thoroughbreds
1940: The Saint's Double Trouble; The Saint Takes Over; Anne of Windy Poplars; Laddie
1941: The Saint in Palm Springs; They Met in Argentina (w/ Leslie Goodwins); Father Takes a Wife; Four Jacks and a Jill
1942: Street of Chance
1948: Are You with It?
1976: Starbird and Sweet William

FREDRIC HOBBS

1973: Alabama's Ghost; Roseland

SANDRA HOCHMAN

1973: Year of the Woman (doc.)

MICHAEL HODGES

1972: Pulp
1974: The Terminal Man

ARTHUR HOERL

1932: Big Town
1933: The Shadow Laughs; Before Morning
1934: Drums o' Voodoo

MICHAEL HOEY

1966: The Navy vs. the Night Monsters

DAVID HOFFMAN

1974: Sing Sing Thanksgiving

HERMAN HOFFMAN

1951: The MGM Story
1955: It's a Dog's Life
1956: The Great American Pastime; The Battle of Gettysburg (doc.)
1957: The Invisible Boy

JOHN HOFFMAN

1945: The Crimson Canary; Strange Confession
1948: The Wreck of the Hesperus
1949: The Lone Wolf and His Lady
1950: I Killed Geronimo

RENAUD HOFFMAN

1930: Blaze o' Glory (w/ J. Crone); The Climax

JAMES HOGAN (1891-1943)

1931: The Sheriff's Secret; Six Shooters in Lariat; Echo of the 45
1936: The Accusing Finger; Arizona Raiders; Desert Gold
1937: Bulldog Drummond Escapes; Ebb Tide; The Last Train from Madrid; Arizona Mahoney
1938: Bulldog Drummond's Peril; The Texans; Scandal Street; Sons of the Legion
1939: Arrest Bulldog Drummond; Bulldog Drummond's Bride; Bulldog Drummond's Secret Police; Grand Jury Secrets; $1,000 a Touchdown
1940: The Farmer's Daughter; Texas Rangers Ride Again; Queen of the Mob
1941: Ellery Queen and the Murder Ring; Ellery Queen and the Perfect Crime; Ellery Queen's Penthouse Mystery; Power Dive
1942: Close Call for Ellery Queen; Desperate Chance for Ellery Queen; Enemy Agents Meet Ellery Queen
1943: The Mad Ghoul; The Strange Death of Adolf Hitler; No Place for a Lady

THEODORE HOLCOMB

1972: Russia (doc.)

LANSING C. HOLDEN

1935: She (w/ Irving Pichel)

WILLIAM HOLE, JR.

1957: Hell Bound
1959: Speed Crazy; The Ghost of Dragstrip Hollow; Four Fast Guns
1962: The Devil's Hand; Twist All Night (The Continental Twist)

ALLEN HOLLEB

1974: Candy Stripe Nurses

BEN HOLMES (-1943)

1934: Lightning Strikes Twice
1936: We're on the Jury; The Plot Thickens; The Farmer in the Dell
1937: There Goes My Girl; Too Many Wives
1938: Maid's Night Out; The Saint in New York; I'm from the City; Little Orphan Annie
1943: Petticoat Larceny

RON HONTHANER

1974: The House on Skull Mountain

RANDALL HOOD

1961: The Two Little Bears

ARTHUR HOPKINS

1933: His Double Life (w/ W. DeMille)

DENNIS HOPPER (1936-)

1969: Easy Rider
1971: The Last Movie

E. MASON HOPPER (1885-1966)

1929: The Carnation Kid; Square Shoulders
1930: Their Own Desire; Temptation; Wise Girls
1932: Her Mad Night; Shop Angel; Midnight Morals; Alias Mary Smith; No Living Witness; Malay Nights
1933: Sister to Judas; One Year Later
1934: Curtain at Eight
1935: Hong Kong Nights

JERRY HOPPER (1907-)

1952: The Atomic City; Hurricane Smith
1953: Pony Express
1954: Alaska Seas; The Secret of the Incas; Naked Alibi
1955: Smoke Signal; The Private War of Major Benson; One Desire; The Square Jungle
1956: Never Say Goodbye; The Toy Tiger; The Sharkfighters; Everything but the Truth
1958: The Missouri Traveler
1961: Blueprint for Robbery
1970: Madron

TOBE HOPPER

1974: The Texas Chainsaw Massacre

LEONARD HORN

1968: Rogue's Gallery

1970: The Magic Garden of Stanley Sweetheart
1972: Corky

JAMES W. HORNE (1880-1942)

1928: The Big Hop
1935: Bonnie Scotland
1936: The Bohemian Girl
1937: Way Out West; All Over Town

(Serials)

1938: The Spider's Web; Flying G-Men (both w/ Ray Taylor)
1940: Deadwood Dick; The Green Archer; The Shadow; Terry and the Pirates
1941: Holt of the Secret Service; The Iron Claw; The Spider Returns; White Eagle
1942: Captain Midnight; Perils of the Royal Mounted

HARRY HORNER (1910-)

1952: Red Planet Mars; Beware, My Lovely
1953: Vicki
1954: New Faces
1955: A Life in the Balance
1956: Man from Del Rio; The Wild Party

ROBERT J. HORNER

1930: The Apache Kid's Escape
1931: Wild West Whoopee; Kid from Arizona
1935: Border Guns; Crimson Trails; Defying the Law; Phantom Cowboy; The Three Renegades; Whirlwind Rider; Western Racketeers
1936: Innocence of the Manhunt; Midnight Secrets; Vice Bondage

IRVING HOROWITZ

1974: Snapshots

JOY N. HOUCK, JR.

1975: Night of the Strangler
1976: The Creature from Black Lake

JOHN HOUGH (1941-) Brit.

(American films only)
1973: The Legend of Hell House
1974: Dirty Mary, Crazy Larry
1975: Escape to Witch Mountain
1978: Return from Witch Mountain; Brass Target

R. L. HOUGH

1932: Silent Witness (w/ Marcel Varnel)

NORMAN HOUSTON

1932: Exposure

CY HOWARD (1915-)

1970: Lovers and Other Strangers
1972: Every Little Crook and Nanny

DAVID HOWARD (1896-1941)

1932: The Rainbow Trail; Mystery Ranch; The Golden West
1933: Smoke Lightning
1934: The Marines Are Coming; Crimson Romance; In Old Santa Fe
1935: Whispering Smith Speaks; Hard Rock Harrigan; Thunder Mountain
1936: O'Malley of the Mounted; Border Patrol; The Mine with the Iron Door; Daniel Boone; Conflict
1937: Park Avenue Logger
1938: The Stadium Murders; Gun Law; Painted Desert; Border G-Man; Lawless Valley; Hollywood Stadium Mystery
1939: The Renegade Ranger; Arizona Legion; The Fighting Gringo; The Marshal of Mesa City; Timber Stampede; Trouble in Sundown
1940: Bullet Code; Legion of the Lawless; Prairie Law; Triple Justice
1941: Dude Cowboy
1942: Six Gun Gold

(Serials)

1933: Mystery Squadron (w/ Colbert Clark)
1934: The Lost Jungle (w/ A. Schaefer)

JAMES HOWARD

1974: Welcome Home, Johnny

RON HOWARD (1954-   )

1977: Grand Theft Auto

SANDY HOWARD

1964: Diary of a Bachelor
1967: One Step to Hell

WILLIAM K. HOWARD (1899-1954)

1929: The Valiant; Love, Live and Laugh; Christina
1930: Scotland Yard; Good Intentions
1931: Don't Bet on Women; Transatlantic; Surrender
1932: The First Year; Sherlock Holmes; The Trial of Vivienne Ware
1933: The Power and the Glory
1934: The Cat and the Fiddle; Evelyn Prentice; This Side of Heaven
1935: Mary Burns, Fugitive; Rendezvous; Vanessa, Her Love Story
1936: The Princess Comes Across; Fire over England (Brit.)
1937: Murder on Diamond Row; Over the Moon (w/ T. Freeland)
1939: Back Door to Heaven
1940: Money and the Woman
1941: Bullets for O'Hara
1942: Klondike Fury
1943: Johnny Come Lately
1944: When the Lights Go On Again
1946: A Guy Could Change

JAMES WONG HOWE (1899-1976)

1954: Go, Man, Go
1958: Invisible Avenger (w/ J. Sledge)

HARRY O. HOYT (1891-1961)

1930: Darkened Skies
1933: Jungle Bride (w/ Albert Kelley)

ROBERT HOYT

1934: Racketeer Round Up

LUCIEN HUBBARD

1929: The Mysterious Island (w/ M. Tourneur, B. Christenson)

RICHARD HUBLER

1947: The Last Nazi?

ROY HUGGINS (1914-)

1952: Hangman's Knot

JOHN HUGH

1955: Yellowneck
1957: Naked in the Sun
1975: The Meal
1979: Deadly Encounter

HOWARD HUGHES (1905-1976)

1930: Hell's Angels
1943: The Outlaw
1950: Vendetta (unc. w/ others)
1957: Jet Pilot (unc. w/ J. von Sternberg)

DON HULETTE

1977: Breaker! Breaker!

H. BRUCE ("LUCKY") HUMBERSTONE (1903-)

1932: Strangers of the Evening; The Crooked Circle; If I Had a Million (w/ J. Cruze, E. Lubitsch, N. Z. McLeod, S. Roberts, W. A. Seiter, N. Taurog)
1933: King of the Jungle (w/ Max Marcin)
1934: The Merry Wives of Reno; Goodbye Love; The Dragon Murder Case
1935: Ladies Love Danger; Silk Hat Kid; Three Live Ghosts
1936: Charlie Chan at the Race Track; Charlie Chan at the Opera
1937: Charlie Chan at the Olympics; Checkers
1938: In Old Chicago (w/ Henry King; Humberstone directed fire sequence); Rascals; Time Out for Murder; Charlie Chan in Honolulu; While New York Sleeps
1939: Pack Up Your Troubles
1940: Lucky Cisco Kid; The Quarterback

1941: Tall, Dark and Handsome; Sun Valley Serenade; I Wake Up Screaming
1942: To the Shores of Tripoli; Iceland
1943: Hello, Frisco, Hello
1944: Pin-Up Girl
1945: Wonder Man; Within These Walls
1946: Three Little Girls in Blue
1947: The Homestretch
1948: Fury at Furnace Creek
1950: South Sea Sinner
1951: Happy Go Lovely
1952: She's Working Her Way Through College
1953: The Desert Song
1955: Ten Wanted Men; The Purple Mask
1957: Tarzan and the Lost Safari (Brit.)
1958: Tarzan's Fight for Life (Brit.)
1962: Madison Avenue

ANDRE HUNEBELLE (1896-)

1966: Shadow of Evil

ED HUNT

1978: Starship Invaders
1979: Alien Encounter

PAUL HUNT

1973: The Clones (w/ Luman Card)
1975: The Great Gundown

PETER H. HUNT (1938-)

1972: 1776
1973: The Homecoming (Brit.)
1976: Shout at the Devil (Brit.)

MAURY HURLEY

1972: It Ain't Easy

HARRY HURWITZ

1971: The Projectionist

LEO HURWITZ (1909-)

1932: Hunger
1934: Scottsboro
1938: Pay Day (w/ P. Strand)
1942: Native Land (w/ P. Strand)
1948: Strange Victory (doc.)

WARIS HUSSEIN

1970: Quackser Fortune Has a Cousin in the Bronx (Ire.)
1972: The Possession of Joel Delaney

JOHN HUSTON (1906-)

1941: The Maltese Falcon
1942: Across the Pacific; In This Our Life
1948: Key Largo; The Treasure of Sierra Madre
1949: We Were Strangers
1950: The Asphalt Jungle
1951: The African Queen; The Red Badge of Courage
1952: Moulin Rouge
1954: Beat the Devil
1956: Moby Dick
1957: Heaven Knows, Mr. Allison
1958: The Barbarian and the Geisha; The Roots of Heaven
1960: The Unforgiven
1961: The Misfits
1962: Freud
1963: The List of Adrian Messenger
1964: The Night of the Iguana
1966: The Bible
1967: Casino Royale (w/ others); Reflections in a Golden Eye
1969: Sinful Davey (Brit.); A Walk with Love and Death
1970: The Kremlin Letter
1972: Fat City; The Life and Times of Judge Roy Bean
1973: The Mackintosh Man
1975: The Man Who Would Be King
1979: Wise Blood

CHARLES HUTCHISON

1931: Women Men Marry; Private Scandal
1932: Bachelor Mother; Out of Singapore
1933: Found Alive
1934: House of Danger; Pals of the Prairie
1935: Circus Shadows; Judgment Book; On Probation; Riddle Ranch
1936: Born to Fight; Desert Guns; Night Cargo; Phantom Patrol
1938: Topa Topa (w/ V. Moore)
1940: Killers of the Wild (w/ V. Moore)

BRIAN G. HUTTON (1935-)

1965: Wild Seed
1966: The Pad (and How to Use It)
1968: Sol Madrid
1969: Where Eagles Dare
1970: Kelly's Heroes
1972: X Y & Zee
1973: Night Watch

ROBERT HUTTON (1920-)

1962: The Slime People

WILLARD HUYCK

1979: French Postcards

PETER HYAMS (1943-)

1974: Busting; Our Time
1976: Peeper
1978: Capricorn One
1979: Hanover Street

RALPH INCE (1887-1937)

1929: Hurricane
1934: Men of America; Lucky Devils; Flaming Gold; No Escape (Brit.); What's a Name (Brit.)
1935: Murder at Monte Carlo (Brit.); Mr. What's His Name (Brit.); Unlimited (Brit.); Black Mask (Brit.); Blue Smoke (Brit.); Rolling House (Brit.)
1936: Jury's Evidence (Brit.); It's You I Want (Brit.); Jail Break (Brit.) 12 Good Men (Brit.); Fair Exchange (Brit.); Hail and Farewell (Brit.)
1937: The Vulture (Brit.); Side Street Angel (Brit.); It's Not Cricket (Brit.); The Perfect Crime (Brit.); The Man Who Made Diamonds (Brit.)

LLOYD INGRAHAM (1885-1956)

1930: Take the Heir

BORIS INGSTER (c. 1913-)

1940: The Stranger on the Third Floor
1949: The Judge Steps Out
1950: Southside 1-1000

STEVE INHAT

1972: The Honkers

JOHN IRELAND (1914-)

1953: Outlaw Territory (w/ Lee Garmes)

JACK IRWIN

1931: Lightnin' Smith's Return (alt. Valley of the Bad Men); White Renegade (orig. The Empire Builders)

ANTHONY ISASI

1966: That Man in Istanbul
1973: Summertime Killer

NEIL ISRAEL

1976: Tunnelvision (w/ B. Swirnoff)
1979: Americathon

JAMES IVORY (1928-)

1969: The Guru (Brit.)
1970: Bombay Talkie (Brit./Ind.)
1972: Savages
1975: The Wild Party; Autobiography of a Princess
1976: Sweet Sounds
1977: Roseland
1979: The Europeans (Brit.)

JACQUES JACCARD

1930: The Cheyenne Kid
1936: Senor Jim
1937: Phantom of Santa Fe

DEL JACK

1969: A Session with the Committee

HORACE JACKSON

1974: Tough

LARRY JACKSON

1975: Bugs Bunny Superstar (anim.)

MIKE JACKSON

1979: The Search for Solutions (doc.)

PATRICK JACKSON (1916-   ) Brit.

(American film only)
1950: Shadow on the Wall

RICHARD JACKSON

1973: The Big Bustout

WILFRED JACKSON

1955: Lady and the Tramp (anim. w/ others)

ARTHUR JACOBSON

1935: Home on the Range

IRVING JACOBY

1954: The Lonely Night (semi-doc.)
1968: Snow Treasure

JOSEPH JACOBY

1973: Hurry Up, or I'll Be 30
1977: The Great Bank Hoax

FELIX JACOVES

1948: Embraceable You
1949: Homicide

HENRY JAGLOM

1971: A Safe Place

ALAN JAMES

1932: Come on, Tarzan; Fargo Express; Tombstone Canyon
1933: Gun Justice; King of the Arena; The Lone Avenger; Phantom Thunderbolt; The Strawberry Roan; Trail Drive
1934: Honor on the Range; Smoking Guns; Wheels of Destiny; When a Man Sees Red
1935: Valley of Wanted Men; Men of Action; Arizona Trails
1936: Lucky Terror; Swifty
1938: Call of the Rockies; Two Gun Justice; West of Rainbow's End; Land of Fighting Men
1939: Trigger Smith
1943: The Law Rides Again; Wild Horse Stampede

(Serials)

1937: Dick Tracy (w/ Ray Taylor); Flaming Frontier (w/ R. Taylor); S.O.S. Coast Guard (w/ W. Witney)
1938: Red Barry (w/ F. Beebe)
1939: Scouts to the Rescue (w/ R. Taylor)

J. FRANK JAMES

1979: The Sweet Creek County War

RIAN JAMES

1933: Best of Enemies

JERRY JAMESON

1972: The Dirt Gang
1974: The Bat People
1977: Airport '77

CHARLES JARROTT (1927-)

1969: Anne of the 1000 Days (Brit.)

1971: Mary, Queen of Scots
1973: Lost Horizon
1974: The Dove (Brit.)
1977: The Other Side of Midnight; The Littlest Horse Thieves

LEIGH JASON (JACOBSON) (1904-   )

1929: Wolves of the City; Eyes of the Underworld; Tip Off
1933: High Gear
1936: Love on a Bet; The Bride Walks Out; That Girl from Paris
1937: Wise Girl
1938: The Mad Miss Manton
1939: The Flying Irishman; Career
1941: Model Wife; Three Girls About Town; Lady for a Night
1943: Dangerous Blondes
1944: Nine Girls; Caroline Blues
1946: Meet Me on Broadway
1947: Lost Honeymoon; Out of the Blue; Man from Texas
1952: Okinawa

WILL JASON (1910-   )

1944: The Soul of a Monster
1945: Eve Knew Her Apples; Tahiti Nights; Ten Cents a Dance
1946: Blonde Alibi; The Dark Horse; Idea Girl; Slightly Scandalous
1947: Sarge Goes to College
1948: Campus Sleuth; Music Man; Rusty Leads the Way
1951: Chain of Circumstance; Disc Jockey
1952: Thief of Damascus

WILFRED JAXON

1951: Alice in Wonderland

GEORGE JESKE

1933: Flaming Signal (w/ C. E. Roberts)

NORMAN JEWISON (1926-   )

1962: Forty Pounds of Trouble
1963: The Thrill of It All
1964: Send Me No Flowers
1965: The Art of Love; The Cincinnati Kid
1966: The Russians Are Coming, the Russians Are Coming
1967: In the Heat of the Night
1968: The Thomas Crown Affair
1969: Gaily, Gaily

1971: Fiddler on the Roof
1973: Jesus Christ, Superstar
1975: Rollerball
1978: F.I.S.T.
1979: ... And Justice for All

CHICK JOHNSON

1973: The Black Moses of Soul (doc.)

EMORY JOHNSON

1930: The Third Alarm
1932: The Phantom Express

LAMONT JOHNSON (1920-)

1967: A Covenant with Death
1968: Kona Coast
1970: My Sweet Charlie; The McKenzie Break (Brit.)
1971: A Gunfight
1972: The Groundstar Conspiracy; You'll Like My Mother
1973: The Last American Hero
1974: Visit to a Chief's Son
1976: Lipstick
1977: One on One
1978: Somebody Killed Her Husband

NUNNALLY JOHNSON (1897-1977)

1954: Night People; Black Widow
1955: How to Be Very, Very Popular
1956: The Man in the Gray Flannel Suit
1957: Oh, Men! Oh, Women!; The Three Faces of Eve
1959: The Man Who Understood Women
1960: The Angel Wore Red (Ital.)

RAYMOND K. JOHNSON

1931: Call of the Rockies
1935: Kentucky Blue Streak; Skybound
1936: I'll Name the Murderer; Suicide Squad
1939: Daughter of the Tong; In Old Montana; Code of the Fearless; Two Gun Troubador
1940: Covered Wagon Trails; The Kid from Santa Fe; Land of Six Guns; Wild Horse Range; The Cheyenne Kid; Pinto Canyon; Riders from Nowhere; Ridin' the Trail
1941: Law of the Wild; Law of the Wolf

TED JOHNSON

1977: Andy Warhol's Bed

BUCK JONES (1889-1942)

1937: Black Aces; For the Service; Law for Tombstone

CHUCK JONES

1979: The Bugs Bunny/Road-Runner Movie (anim.)

EUGENE S. JONES

1968: A Face of War (doc.)

F. RICHARD JONES

1929: Bulldog Drummond

GROVER JONES

1933: Hell and High Water (w/ William Slavens McNutt)

HARDY JONES

1979: Dolphin (doc. w/ Michael Wiese)

HARMON C. JONES (1911-1972)

1951: As Young As You Feel
1952: The Pride of St. Louis; Bloodhounds of Broadway
1953: The Silver Ship; City of Bad Men; The Kid from Left Field
1954: Gorilla at Large; Princess of the Nile
1955: Target Zero
1956: A Day of Fury; Canyon River
1958: The Beast of Budapest; Bullwhip; Wolf Larsen
1966: Don't Worry, We'll Think of a Title

L. Q. JONES

1975: A Boy and His Dog

WINSTON JONES

1956: UFO (Unidentified Flying Objects) (doc.)

RON JOY

1971: The Animals

JOEL JUDGE (see ABNER BIBERMAN)

RUPERT JULIAN (1889-1943)

1930: Love Comes Along; The Cat Creeps

NATHAN (HERTZ) JURAN (1907-)

1952: The Black Castle
1953: Gunsmoke; Law and Order; The Golden Blade; Tumbleweed
1954: Highway Dragnet; Drums Across the River
1955: The Crooked Web
1957: The Deadly Mantis; Hellcats of the Navy; Twenty Million Miles to Earth
1958: The Seventh Voyage of Sinbad; Good Day for a Hanging; Attack of the 50-foot Woman
1961: Flight of the Lost Balloon
1962: Jack the Giant Killer
1963: Siege of the Saxons
1964: First Men In the Moon
1965: East of Sudan (Brit.)
1970: Land Raiders (Span.)
1973: The Boy Who Cried Werewolf

JON JUST

1977: Angel City

JAN KADAR (1918-1979) Czech.

(American films only)
1970: The Angel Levine
1978: Freedom Road

ELLIS KADISON

1965: Git
1966: The Cat

JEREMY PAUL KAGAN

1977: Heroes; Scott Joplin
1978: The Big Fix

RICHARD C. KAHN

1931: Secret Menace
1934: Children of Loneliness
1939: The Bronze Buckaroo; Harlem Rides Again; Two-Gun Man from Harlem
1940: Son of Ingagi (w/ Herbert Meyer); Buzzy Rides the Range (alt. Western Terror)
1941: Buzzy and the Phantom Pinto

JOSEPH KANE (1897-)

1935: Tumbling Tumbleweeds; Melody Trail; The Sagebrush Troubadour
1936: The Lawless Nineties; King of the Pecos; The Lonely Trail; Guns and Guitars; Oh, Susanna!; Ride, Ranger, Ride
1937: Paradise Express; Git Along Little Dogies; Ghost Town Gold; Round-Up Time in Texas; The Old Corral; Come On, Cowboys!; Gunsmoke Ranch; Public Cowboy No. One; Yodelin' Kid from Pine Ridge; Boots and Saddles; Springtime in the Rockies
1938: The Old Barn Dance; Born to Be Wild; Arson Gang Busters; Under Western Skies; Gold Mine in the Sky; Man from Music Mountain; Billy the Kid Returns; Come on, Rangers!; Shine on Harvest Moon
1939: Rough Riders' Round-Up; Frontier Pony Express; Southward Ho!; In Old Caliente; In Old Monterey; Wall Street Cowboy; The Arizona Kid; Days of Jesse James; Saga of Death Valley
1940: Young Buffalo Bill; The Carson City Kid; The Ranger and the Lady; Colorado; Young Bill Hickok; The Border Legion
1941: Robin Hood of the Pecos; In Old Cheyenne; Sheriff of Tombstone; The Great Train Robbery; Nevada City; Rags to Riches; Bad Man of Deadwood; Jesse James at Bay; Red River Valley
1942: The Man from Cheyenne; South of Santa Fe; Sunset on the Desert; Romance of the Range; Sons of the Pioneers; Sunset Serenade; Heart of the Golden West; Ridin' Down the Canyon
1943: Idaho; King of the Cowboys; Song of Texas; Silver Spurs; The Man from Music Mountain; Hands Across the Border
1944: The Cowboy and the Senorita; The Yellow Rose of Texas; Song of Nevada

1945: Flame of the Barbary Coast; The Cheaters; Dakota
1946: In Old Sacramento; The Plainsman and the Lady
1947: Wyoming
1948: Old Los Angeles; The Gallant Legion; The Plunderers
1949: The Last Bandit; Brimstone
1950: Rock Island Trail; The Savage Horde; California Passage
1951: Oh! Susanna; Fighting Coast Guard; The Sea Hornet
1952: Hoodlum Empire; Woman of the North Country; Ride the Man Down
1953: San Antone; Fair Wind to Java; Sea of Lost Ships
1954: Jubilee Trail
1955: Hell's Outpost; Timberjack; The Road to Denver; The Vanishing American
1956: The Maverick Queen; Thunder over Arizona; Accused of Murder
1957: Duel at Apache Wells, Spoilers of the Forest; Last Stagecoach West; The Crooked Circle
1958: Gunfire at Indian Gap; The Notorious Mr. Monks; The Lawless Eighties; The Man Who Died Twice
1966: Here Comes That Nashville Sound (Country Boy)
1967: Search for the Evil One
1968: Track of Thunder
1971: Smoke in the Wind (w/ unc. Andy Brennan; unreleased)

(Serials)

1935: Fighting Marines (w/ B. R. Eason)
1936: Darkest Africa; The Undersea Kingdom (both w/ B. R. Eason)
1949: King of the Jungleland (w/ B. R. Eason)

JEFF KANEW

1972: Black Rodeo
1979: Natural Enemies

GARSON KANIN (1912-)

1938: A Man to Remember; Next Time I Marry
1939: Bachelor Mother; The Great Man Votes
1940: My Favorite Wife; They Knew What They Wanted
1941: Tom, Dick and Harry
1945: The True Glory (doc. w/ Carol Reed)
1969: Some Kind of a Nut; Where It's At

MICHAEL KANIN (1910-)

1951: When I Grow Up

HAL KANTER (1918-   )

1957: Loving You
1958: I Married a Woman;   Once Upon a Horse
1963: Hot Horse

JONATHAN KAPLAN (1947-   )

1973: The Slams; The Student Teachers
1974: Truck Turner; Night Call Nurses
1975: White Line Fever
1977: Mr. Billion
1979: Over the Edge

RICHARD KAPLAN

1965: The Eleanor Roosevelt Story (doc.)
1979: A Look at Liv (doc.)

LESLIE KARDOS

1940: Dark Streets of Cairo
1951: The Strip
1953: Small Town Girl
1957: The Man Who Turned to Stone; The Tijuana Story

PETER KARES

1975: The Night They Robbed Big Bertha's

PHIL KARLSON (KARLSTEIN) (1908-   )

1944: A Wave, a Wac and a Marine
1945: G.I. Honeymoon; There Goes Kelly; The Shanghai Cobra
1946: Live Wires; Dark Alibi; Behind the Mask; Bowery Bombshell; The Missing Lady; Wife Wanted
1947: Black Gold; Kilroy Was Here; Louisiana
1948: Rocky; Adventure in Silverado; Thunderhoof
1949: Ladies of the Chorus; The Big Cat; Down Memory Lane
1950: The Iroquois Trail
1951: Lorna Doone; The Texas Rangers; Mask of the Avenger
1952: Scandal Sheet; The Brigand; Kansas City Confidential
1953: River Street
1954: They Rode West
1955: Tight Spot; Hell's Island; Five Against the House; The Phenix

City Story
1957: The Brothers Rico
1958: Gunman's Walk
1960: Hell to Eternity; Key Witness
1961: The Secret Ways; The Young Doctors
1962: Kid Galahad
1963: Rampage
1966: The Silencers
1967: A Time for Killing
1969: The Wrecking Crew
1970: Hornets' Nest (Ital.)
1972: Ben
1973: Walking Tall
1975: Framed

BILL KARN

1955: Gang Busters
1960: Ma Barker's Killer Brood

ERIC KARSON

1979: Dirt (w/ Cal Naylor)

LEONARD KASTLE

1970: The Honeymoon Killers

MILTON KATSELAS

1972: Butterflies Are Free
1973: Forty Carats
1975: Report to the Commissioner
1979: When You Comin' Back, Red Ryder?

GLORIA KATZ

1975: Messiah of Evil

LEE H. KATZIN

1967: Hondo and the Apaches
1969: Heaven With a Gun; Whatever Happened to Aunt Alice?
1970: The Phynx
1971: Le Mans
1972: The Salzburg Connection

SAM KATZMAN

1937: Orphan of the Pecos; Brothers of the West; Lost Ranch

GEORGE S. KAUFMAN (1889-1961)

1947: The Senator Was Indiscreet

LLOYD KAUFMAN

1971: The Battle of Love's Return

MILLARD KAUFMAN

1962: Convicts Four (orig. Reprieve)

PHILIP KAUFMAN (1936-)

1965: Goldstein (w/ Benjamin Monaster)
1969: Fearless Frank
1972: The Great Northfield Minnesota Raid
1974: The White Dawn
1978: Invasion of the Body Snatchers
1979: The Wanderers; Up Your Ladder

HELMUT KAUTNER (1908-) Germ.

(American films only)
1958: The Restless Years
1959: Stranger in My Arms

ROGER KAY

1962: The Cabinet of Dr. Caligari

GILBERT L. KAY (Brit.)

(American films only)
1956: Three Bad Sisters
1964: The Secret Door

EDWARD E. KAYE

1941: Escort Girl

ROBERT KAYLOR

1971: Derby (doc.)
1974: The Nine Lives of Fritz the Cat (anim.)

ELIA KAZAN (1909-)

1945: A Tree Grows in Brooklyn
1946: The Sea of Grass
1947: Boomerang; Gentleman's Agreement
1949: Pinky
1950: Panic in the Streets
1951: A Streetcar Named Desire
1952: Viva Zapata!
1953: Man on a Tightrope
1954: On the Waterfront
1955: East of Eden
1956: Baby Doll
1957: A Face in the Crowd
1960: Wild River
1961: Splendor in the Grass
1963: America, America
1969: The Arrangement
1972: The Visitors
1976: The Last Tycoon

VERNON KEAYS

1942: Strictly in the Groove; Arizona Trail
1944: Trigger Law; The Utah Kid; Marshal of Gunsmoke; Trail to Gunsight
1945: Rockin' in the Rockies; Blazing the Western Trail; Lawless Empire; Dangerous Intruder
1946: Landrush
1948: Whirlwind Raiders

(Serials)

1946: Mysterious Mr. M (w/ Lew Collins)

WILLIAM KEIGHLEY (1889-)

1932: The Match King (w/ Howard Bretherton)
1933: Ladies They Talk About (w/ Howard Bretherton)
1934: Easy to Love; Journal of a Crime; Dr. Monica; Kansas City Princess; Big Hearted Herbert; Babbitt
1935: The Right to Live; G-Men; Mary Jane's Pa; Special Agent; Stars over Broadway
1936: The Singing Kid; Bullets or Ballots; The Green Pastures (w/ Marc Connelly); God's Country and the Woman

1937: The Prince and the Pauper; Varsity Show
1938: The Adventures of Robin Hood (w/ Michael Curtiz); Brother Rat; Secrets of an Actress
1939: Each Dawn I Die; Yes, My Darling Daughter
1940: The Fighting 69th; Torrid Zone; No Time for Comedy
1941: Four Mothers; The Bride Came C.O.D.
1942: George Washington Slept Here; The Man Who Came to Dinner
1947: Honeymoon
1948: The Street with No Name
1950: Rocky Mountain
1951: Close to My Heart
1953: The Master of Ballantrae

HARRY KELLER (1913-)

1949: The Blonde Bandit
1950: Tarnished
1951: Fort Dodge Stampede; Desert of Lost Men
1952: Rose of Cimarron; Leadville Gunslinger; Black Hills Ambush; Thundering Caravans
1953: Marshal of Cedar Rock; Savage Frontier; Bandits of the West; El Paso Stampede
1954: Red River Shore; Phantom Stallion
1956: The Unguarded Moment
1957: Man Afraid; Quantez
1958: The Day of the Bad Man; The Female Animal; Voice in the Mirror; Step Down to Terror
1960: Seven Ways from Sundown
1961: Tammy Tell Me True
1962: Six Black Horses
1963: Tammy and the Doctor
1964: The Brass Bottle
1968: In Enemy Country

ALBERT KELLEY

1930: Woman Racket (w/ Robert Ober)
1933: Jungle Bride (w/ Harry O. Hoyt)
1941: Double Cross
1943: Submarine Base
1948: Street Corner; Slippery McGee

(Serials)

1930: The Leather Pushers

ROY KELLINO

1939: I Met a Murderer
1952: Lady Possessed (w/ W. Spier)
1957: The Silken Affair

BOB KELLJAN (also KELLJCHIAN)

1970: Count Yorga, Vampire
1971: Return of Count Yorga
1973: Scream Blacula Scream
1974: Act of Vengeance
1977: Black Oak Conspiracy

RAY KELLOGG

1959: The Giant Gila Monster; The Killer Shrews
1960: My Dog Buddy
1968: The Green Berets (w/ John Wayne)

DUKE KELLY

1976: My Name Is Legend

GENE KELLY (1912-   )

1949: On the Town (w/ S. Donen)
1952: Singin' in the Rain (w/ S. Donen)
1955: It's Always Fair Weather (w/ S. Donen)
1956: Invitation to the Dance
1957: The Happy Road
1958: The Tunnel of Love
1962: Gigot
1967: A Guide for the Married Man
1969: Hello, Dolly!
1970: The Cheyenne Social Club
1976: That's Entertainment, Part 2

RON KELLY

1970: King of the Grizzlies

JACK KEMP

1948: Miracle in Harlem

BURT KENNEDY (1923-   )

1961: The Canadians
1964: Mail Order Bride
1965: The Rounders
1966: The Money Trap; Return of the Seven
1967: The War Wagon; Welcome to Hard Times

1969: The Good Guys and the Bad Guys; Support Your Local Sheriff; Young Billy Young
1970: Dirty Dingus Magee
1971: The Deserter; Hannie Caulder (Brit.); Support Your Local Gunfighter
1973: The Train Robbers
1976: The Killer Inside Me

WILLIS KENT

1940: Mad Youth

ERLE C. KENTON (1896-)

1929: Trial Marriage; Father and Son; Song of Love
1930: Mexicali Rose; A Royal Romance
1931: The Last Parade; Lover Come Back; Leftover Ladies; X Marks the Spot
1932: Stranger in Town; Guilty as Hell
1933: Island of Lost Souls; From Hell to Heaven; Disgraced!; Big Executive
1934: Search for Beauty; You're Telling Me
1935: The Best Man Wins (w/ E. R. Davison); Party Wire; The Public Menace; Grand Exit
1936: Devil's Squadron; Counterfeit; End of the Trail
1937: Devil's Playground; Racketeers in Exile; She Asked for It
1938: The Lady Objects; Little Tough Guys in Society
1939: Everything's on Ice; Escape to Paradise
1940: Remedy for Riches
1941: Petticoat Politics; Melody for Three; Naval Academy; They Meet Again; Flying Cadets
1942: Frisco Lil; North to the Klondike; The Ghost of Frankenstein; Pardon My Sarong; Who Done It?
1943: How's About It?; It Ain't Hay; Always a Bridesmaid
1945: House of Frankenstein; She Gets Her Man; House of Dracula
1946: The Cat Creeps; Little Miss Big
1948: Bob and Sally
1950: One Too Many

PAUL KENWORTHY, JR.

1957: Perri (w/ Ralph Wright)

JAMES V. KERN (1909-1966)

1944: The Doughgirls
1946: Never Say Goodbye
1947: Stallion Road
1948: April Showers
1951: The Second Woman; Two Tickets to Broadway

SARAH KERNOCHAN

1972: Marjoe (doc. w/ Howard Smith)

GLENN KERSHNER

1937: Island Captives

IRVIN KERSHNER (1923-)

1958: Stakeout on Dope Street
1959: The Young Captives
1961: The Hoodlum Priest
1963: A Face in the Rain
1964: The Luck of Ginger Coffey (Can.)
1966: A Fine Madness
1967: The Flim Flam Man
1970: Loving
1972: Up the Sandbox
1974: S*P*Y*S*
1976: Return of a Man Called Horse
1978: The Eyes of Laura Mars

HARRY E. KERWIN

1975: God's Bloody Acre
1979: Barracuda

HENRY S. KESLER

1943: Three Russian Girls (w/ Fedor Ozep)
1957: Five Steps to Danger

BRUCE KESSLER

1968: Angels from Hell; Killers Three
1969: The Gay Deceivers
1971: Simon, King of the Witches

ROLAND KIBBEE

1974: The Midnight Man

MICHAEL KIDD (1919-)

1958: Merry Andrew

EDWARD KILLY

1935: Freckles (w/ W. Hamilton); Seven Keys to Baldpate (w/ W. Hamilton)
1936: Murder on a Bridle Path; Bunker Bean (both w/ W. Hamilton); Second Wife; Wanted-Jane Turner
1937: China Passage; The Big Shot; Saturday's Heroes
1938: Quick Money
1940: The Fargo Kid; Stage to Chino; Wagon Train
1941: Along the Rio Grande; The Bandit Trail; Cyclone on Horseback; Robbers of the Range
1942: Come On, Danger; Land of the Open Range; Riding the Wind
1944: Nevada
1945: West of the Pecos; Wanderer of the Wasteland (w/ W. Grissell)

BILL KIMBERLIN

1979: American Nitro

CLINTON KIMBRO

1973: The Young Nurses

ALLAN KING (1930-) Can.

(American films only)
1969: A Married Couple

BURTON KING

1929: In Old California

HENRY KING (1888-)

1929: She Goes to War
1930: Hell Harbor; Eyes of the World; Lightnin'; The Concentratin' Kid
1931: Merely Mary Ann; Over the Hill
1932: The Woman in Room 13
1933: State Fair; I Loved You Wednesday (w/ W. C. Menzies)
1934: Carolina; Marie Galante
1935: Way Down East; One More Spring
1936: Lloyds of London; Ramona; The Country Doctor
1937: Seventh Heaven
1938: Alexander's Ragtime Band; In Old Chicago
1939: Jesse James; Stanley and Livingstone
1940: Chad Hanna; Little Old New York; Maryland

1941: Remember the Day; A Yank in the RAF
1942: The Black Swan
1943: The Song of Bernadette
1944: Wilson
1945: A Bell for Adano
1946: Margie
1947: Captain from Castile
1948: Deep Waters
1949: Prince of Foxes; 12 O'Clock High
1950: The Gunfighter
1951: David and Bathsheba; I'd Climb the Highest Mountain
1952: The Snows of Kilimanjaro; Wait 'Til the Sun Shines, Nellie; O. Henry's Full House (w/ others)
1953: King of the Khyber Rifles
1955: Love Is a Many Splendored Thing; Untamed
1956: Carousel
1957: The Sun Also Rises
1958: The Bravados; The Old Man and the Sea (unc. w/ others)
1959: Beloved Infidel; This Earth Is Mine
1962: Tender Is the Night

LOUIS KING (1898-1962)

1929: The Vagabond Cub; The Freckled Rascal; The Little Savage; Pals of the Prairie
1930: The Lone Rider; Shadow Ranch; Men Without Law
1931: Desert Vengeance; The Fighting Sheriff; Border Law; The Deceiver
1932: Police Court; The Country Fair; Arm of the Law; Drifting Souls
1933: Robbers' Roost; Life in the Raw
1934: Murder in Trinidad; Pursued; Bachelor of Arts
1935: Charlie Chan in Egypt
1936: Road Gang; Special Investigator; Song of the Saddle; The Bengal Tiger
1937: Melody for Two; That Man's Here Again; Draegerman Courage; Wild Money; Bulldog Drummond Comes Back; Wine, Women and Horses; Bulldog Drummond's Revenge
1938: Tip-Off Girls; Hunted Men; Bulldog Drummond in Africa; Illegal Traffic; Tom Sawyer, Detective
1939: Persons in Hiding; Undercover Doctor
1940: Seventeen; Typhoon; The Way of All Flesh; Moon over Burma
1942: Young America
1943: Chetniks
1944: Ladies of Washington
1945: Thunderhead--Son of Flicka
1946: Smoky
1947: Thunder in the Valley
1948: Green Grass of Wyoming
1949: Sand; Mrs. Mike
1950: Frenchie
1952: The Lion and the Horse

1953: Powder River; Sabre Jet
1954: Dangerous Mission
1956: Massacre

RICK KING

1977: Off the Wall

WOODIE KING

1976: The Long Night

JACK KINNEY

1969: 1001 Arabian Nights (anim.)

DAVID KIRKLAND

1931: Riders of the Cactus
1932: Soul of Mexico

JOHN KIRKLAND

1974: Curse of the Headless Horseman

RAY KIRKWOOD

1935: The Shadow of Silk Lennox

JOHN KIRSCH

1979: Jesus (w/ Peter Sykes)

LEONARD KIRTMAN

1976: Carnival of Blood

ALF KJELLIN (1920-   ) Swed.

(American films only)
1969: Midas Run
1970: The McMasters

ROBERT KLANE

1978: Thank God It's Friday

CHARLES KLEIN (1898-) Germ.

1928: Blindfold
1929: The Sin Sister; Pleasure Crazed

LARRY KLEIN

1970: The Adversary

WILLIAM KLEIN (1926-)

1970: Mr. Freedom; Eldridge Cleaver (doc.); Float Like a Butterfly, Sting Like a Bee

RANDAL KLEISER

1978: Grease

BENJAMIN KLINE

1943: Cowboy in the Clouds
1944: Cowboy from Lonesome River; Cyclone Prairie Rangers; Saddle Leather Law; Sagebrush Heroes; Sundown Valley

(Serials)

1931: Lightning Warrior (w/ A. Schaefer)

HERBERT KLINE (1909-)

1937: Heart of Spain (doc.)
1938: Crisis
1940: Lights Out in Europe (Brit. doc.)
1941: The Forgotten Village (Mex. doc.)
1947: My Father's House (Isr.)
1949: The Kid from Cleveland
1952: The Fighter
1976: The Challenge

LARRY KLINGMAN

1974: Dreams and Nightmares (doc.)

CHRISTOPHER C. KNIGHT

1970: Carry It On (doc.)

EDWIN H. KNOPF (1899-)

1930: Slightly Scarlet (w/ Louis Gasnier); The Light of Western Stars (w/ Otto Brower); Paramount on Parade (w/ others); Border Legion (w/ O. Brower); The Santa Fe Trail (w/ O. Brower); Only Saps Work (w/ Cyril Gardner)
1932: Nice Women
1951: The Law and the Lady

HOWARD W. KOCH (1916-)

1954: Shield for Murder (w/ Edmond O'Brien)
1955: Big House, U.S.A.
1957: Untamed Youth; Bop Girl; Jungle Heat; The Girl in Black Stockings
1958: Fort Bowie; Violent Road; Frankenstein--1970; Andy Hardy Comes Home
1959: The Last Mile; Born Reckless
1973: Badge 373

PANCHO KOHNER

1974: Mr. Sycamore

BARBARA KOPPLE

1977: Harlan County, U.S.A.

ALEXANDER KORDA (1893-1956) Brit.

(American films only)
1929: The Squall; Love and the Devil; Her Private Life
1930: Lilies of the Field; The Princess and the Plumber; Women Everywhere

ZOLTAN KORDA (1895-1961)

1942: The Jungle Book
1943: Sahara
1945: Counter-Attack
1947: The Macomber Affair
1948: A Woman's Revenge
1952: African Fury

JOHN KORTY (1936-   )

1966: The Crazy Quilt
1967: Funnyman
1970: Riverrun
1974: Silence
1976: Alex and the Gypsy
1978: Oliver's Story

HENRY KOSTER (HERMANN KOSTERLITZ) (1905-   )

1935: Peter (Hung.)
1937: Three Smart Girls; One Hundred Men and a Girl
1938: The Rage of Paris
1939: Three Smart Girls Grow Up; First Love
1940: Spring Parade
1941: It Started with Eve
1942: Between Us Girls
1944: Music for Millions
1946: Two Sisters from Boston
1947: The Unfinished Dance; The Bishop's Wife
1948: The Luck of the Irish
1949: Come to the Stable; The Inspector General
1950: Wabash Avenue; My Blue Heaven; Harvey
1951: No Highway in the Sky; Mr. Belvedere Rings the Bell; Elopement
1952: O. Henry's Full House (w/ others); Stars and Stripes Forever; My Cousin Rachel
1953: The Robe
1954: Desiree
1955: A Man Called Peter; The Virgin Queen; Good Morning Miss Dove
1956: D-Day, the Sixth of June; The Power and the Prize
1957: My Man Godfrey
1958: Fraulein
1959: The Naked Maja (Ital.)
1960: The Story of Ruth
1961: Flower Drum Song
1962: Mr. Hobbs Takes a Vacation
1963: Take Her, She's Mine
1965: Dear Brigitte
1966: The Singing Nun

TED KOTCHEFF (1931-   )

1963: Tiara Tahiti
1974: Billy Two Hats (ret. The Lady and the Outlaw); The Apprenticeship of Duddy Kravitz (Can.)
1977: Fun with Dick and Jane
1978: Who Is Killing the Great Chefs of Europe?
1979: North Dallas Forty

YAPHET KOTTO (1937-)

1972: The Limit

BERNARD KOWALSKI (c. 1933-)

1958: Hot Car Girl; Night of the Blood Beast
1959: Attack of the Giant Leeches; Blood and Steel
1969: Krakatoa, East of Java; Stiletto
1970: Macho Callahan
1973: Ssssssss

DAVID KRAMARSKY

1956: The Beast with a Million Eyes

REMI KRAMER

1977: High Velocity

ROBERT KRAMER

1968: The Edge
1970: Ice
1975: Milestones

STANLEY KRAMER (1913-)

1955: Not As a Stranger
1957: The Pride and the Passion
1958: The Defiant Ones
1959: On the Beach
1960: Inherit the Wind
1961: Judgment at Nuremberg
1963: It's a Mad Mad Mad Mad World
1965: Ship of Fools
1967: Guess Who's Coming to Dinner?
1969: The Secret of Santa Vittoria
1970: R.P.M.
1972: Bless the Beasts and Children
1973: Oklahoma Crude
1977: The Domino Principle
1979: The Runner Stumbles

NORMAN KRASNA (1909-)

1943: Princess O'Rourke

1950: The Big Hangover
1956: The Ambassador's Daughter

PAUL KRASNY

1976: Joe Panther

HAROLD F. KRESS (1913-)

1951: No Questions Asked; The Painted Hills
1952: Apache War Smoke

NATHAN KROLL

1964: The Guns of August (doc.)
1974: Helen Hayes; Portrait of an American Actress (doc.)

STANLEY KUBRICK (1928-)

1953: Fear and Desire
1955: Killer's Kiss
1956: The Killing
1957: Paths of Glory
1960: Spartacus
1962: Lolita (Brit.)
1964: Dr. Strangelove (Brit.)
1968: 2001: A Space Odyssey (Brit.)
1971: A Clockwork Orange (Brit.)
1975: Barry Lyndon (Brit.)

BUZZ KULIK (1922-)

1961: The Explosive Generation
1963: The Yellow Canary
1964: Ready for the People
1967: Warning Shot
1968: Sergeant Ryker; Villa Rides; Riot
1972: To Find a Man
1973: Shamus

EDWARD KULL

1935: Man's Best Friend

(Serials)

1935: The New Adventures of Tarzan

PAUL KYRIAZI

1976: Death Machines

GREGORY LA CAVA (1892-1952)

1929: Big News
1930: His First Command
1931: Laugh and Get Rich; Smart Woman
1932: The Age of Consent; The Half-Naked Truth; Symphony of Six Million
1933: Gabriel over the White House; Bed of Roses; Gallant Lady
1934: Affairs of Cellini; What Every Woman Knows
1935: Private Worlds; She Married Her Boss
1936: My Man Godfrey
1937: Stage Door
1939: Fifth Avenue Girl
1940: The Primrose Path
1941: Unfinished Business
1942: Lady in a Jam
1947: Living in a Big Way

HARRY LACHMAN (1886-1975)

1929: Week-End Wives (Brit.); Under the Greenwood Tree (Brit.)
1930: Song of Soho (Brit.); The Yellow Mask (Brit.)
1931: The Love Habit (Brit.)
1932: Aren't We All? (Brit.)
1933: The Face in the Sky; The Outsider (Brit.); Paddy, the Next Best Thing
1934: George White's Scandals (w/ George White, Thornton Freeland); I Like It That Way; Baby Take a Bow
1935: Dante's Inferno; Dressed to Thrill
1936: Charlie Chan at the Circus; Our Relations; The Man Who Lived Twice
1937: The Devil Is Driving; It Happened in Hollywood
1938: No Time to Marry
1940: They Came by Night (Brit.); Murder over New York
1941: Dead Men Tell; Charlie Chan in Rio
1942: Castle in the Desert; The Loves of Edgar Allan Poe; Dr. Renault's Secret

EDWARD LAEMMLE

1929: The Drake Case
1931: Lasca of the Rio Grande
1932: Texas Bad Man
1934: Embarrassing Moments
1935: A Notorious Gentleman

ERNST LAEMMLE

1930: What Men Want

FERNANDO LAMAS (1915-)

1967: The Violent Ones

ANDE LAMB

1950: The Texan Meets Calamity Jane

CHARLES FRED LAMONT (1898-)

1934: The Curtain Falls
1935: Tomorrow's Youth; The World Accuses; Son of Steel; False Pretenses; Gigolette; A Shot in the Dark; Circumstantial Evidence; The Girl Who Came Back; Happiness C.O.D.; The Lady in Scarlet
1936: Ring Around the Moon; Little Red Schoolhouse; Below the Deadline; August Week-End; The Dark Hour; Lady Luck; Bulldog Edition
1937: Wallaby Jim of the Island
1938: International Crime; Shadows over Shanghai; Slander House; Cipher Bureau; The Long Shot
1939: Pride of the Navy; Panama Patrol; Inside Information; Unexpected Father; Little Accident
1940: Oh, Johnny, How You Can Love!; Sandy Is a Lady; Love, Honor and Oh-Baby!; Give Us Wings
1941: San Antonio Rose; Sing Another Chorus; Moonlight in Hawaii; Melody Lane; Road Agent
1942: Don't Get Personal; You're Telling Me; Almost Married; Hi, Neighbor; Get Hep to Love; When Johnny Comes Marching Home
1943: It Comes up Love; Mr. Big; Hit the Ice; Fired Wife; Top Man
1944: Chip off the Old Block; Her Primitive Man; The Merry Monahans; Bowery to Broadway
1945: Salome, Where She Danced; That's the Spirit; Frontier Gal
1946: She Wrote the Book; The Runaround
1947: Slave Girl
1948: The Untamed Breed
1949: Ma and Pa Kettle; Bagdad
1950: Ma and Pa Kettle Go to Town; I Was a Shoplifter; Curtain Call at Cactus Creek; Abbott and Costello in the Foreign Legion
1951: Abbott and Costello Meet the Invisible Man; Comin' Round the Mountain; Flame of Araby
1952: Abbott and Costello Meet Captain Kidd

1953: Abbott and Costello Go to Mars; Ma and Pa Kettle on Vacation; Abbott and Costello Meet Dr. Jekyll and Mr. Hyde
1954: Ma and Pa Kettle at Home; Untamed Heiress; Ricochet Romance
1955: Carolina Cannonball; Abbott and Costello Meet the Keystone Kops; Abbott and Costello Meet the Mummy; Lay That Rifle Down
1956: The Kettles in the Ozarks; Francis in the Haunted House

BURT LANCASTER (1913-)

1955: The Kentuckian

SAUL LANDAU

1971: Fidel (doc.); Brazil: A Report on Torture (doc.)

LEW LANDERS (LOUIS FRIEDLANDER) (1901-1962)

1935: The Raven; Stormy
1936: Parole!; Without Orders; Night Waitress
1937: They Wanted to Marry; The Man Who Found Himself; You Can't Buy Luck; Border Cafe; Flight from Glory; Living on Love; Danger Patrol
1938: Crashing Hollywood; Double Danger; Condemned Women; Law of the Underworld; Blind Alibi; Sky Giants; Smashing the Rackets; Annabel Takes a Tour
1939: Pacific Liner; Twelve Crowded Hours; Fixer Dugan; The Girl and the Gambler; Bad Lands; Conspiracy
1940: Honeymoon Deferred; Enemy Agent; La Conga Nights; Ski Patrol; Wagons Westward; Sing, Dance, Plenty Hot; Girl from Havana; Slightly Tempted
1941: Ridin' on a Rainbow; Lucky Devils; Back in the Saddle; The Singing Hill; I Was a Prisoner on Devil's Island; Mystery Ship; The Stork Pays Off
1942: The Man Who Returned to Life; Alias Boston Blackie; Canal Zone; Harvard, Here I Come; Not a Ladies' Man; Submarine Raider; Cadets on Parade; Atlantic Convoy; Sabotage Squad; The Boogie Man Will Get You; Smith of Minnesota; Stand by All Networks; Junior Army
1943: After Midnight with Boston Blackie; Redhead from Manhattan; Murder in Times Square; Power of the Press; Doughboys in Ireland; Deerslayer
1944: Cowboy Canteen; The Ghost That Walks Alone; The Return of the Vampire; Two-Man Submarine; Stars on Parade; The Black Parachute; U-Boat Prisoner; Swing in the Saddle; I'm from Arkansas
1945: Crime, Inc.; The Power of the Whistler; Trouble Chasers; Follow That Woman; Arson Squad; Shadow of Terror; The Enchanted Forest; Tokyo Rose

1946: The Mask of Dijon; A Close Call for Boston Blackie; The Truth About Murder; Death Valley
1947: Danger Street; Seven Keys to Baldpate; Under the Tonto Rim; Thunder Mountain; The Son of Rusty; Devil Ship
1948: My Dog Rusty; Adventures of Gallant Bess; Inner Sanctum
1949: Stagecoach Kid; Law of the Barbary Coast; Air Hostess; Barbary Pirate
1950: Davy Crockett, Indian Scout; Girls' School; Dynamite Pass; Tyrant of the Sea; State Penitentiary; Beauty on Parade; Chain Gang; Last of the Buccaneers; Revenue Agent
1951: Blue Blood; A Yank in Korea; When the Redskins Rode; The Big Gusher; Hurricane Island; The Magic Carpet; Jungle Manhunt
1952: Aladdin and His Lamp; Jungle Jim in the Forbidden Land; California Conquest; Arctic Flight
1953: Torpedo Alley; Tangier Incident; Man in the Dark; Run for the Hills; Captain John Smith and Pocahontas
1954: Captain Kidd and the Slave Girl
1956: The Cruel Tower
1958: Hot Rod Gang
1963: Terrified

(Serials)

1934: The Red Rider; Tailspin Tommy; The Vanishing Shadow
1935: The Call of the Savage; Rustlers of Red Dog

JAMES LANDIS

1964: The Nasty Rabbit
1971: Deadwood '76

JOHN LANDIS

1973: Schlock
1977: The Kentucky Fried Movie
1978: National Lampoon's Animal House

PAUL LANDRES (1912-)

1949: Grand Canyon; Square Dance Jubilee
1950: Hollywood Varieties; A Modern Marriage
1951: Rhythm Inn; Navy Bound
1952: Army Bound
1953: Eyes of the Jungle
1957: Chain of Evidence; Hell Canyon Outlaws; Last of the Badmen; New Day at Sundown; Oregon Passage; The Vampire; The Return of Dracula
1958: The Flame Barrier; Frontier Gun; Johnny Rocco; Man from God's Country

1959: Lone Texan; Miracle of the Hills
1966: Son of a Gunfighter (Span.)

SIDNEY LANFIELD (1900-1972)

1930: Cheer Up and Smile
1931: Three Girls Lost; Hush Money
1932: Dance Team; Society Girl; Hat Check Girl
1933: Broadway Bad
1934: Moulin Rouge; The Last Gentleman
1935: Hold 'Em, Yale; Red Salute; King of Burlesque
1936: Half Angel; Sing, Baby, Sing; One in a Million
1937: Wake Up and Live; Thin Ice; Love and Hisses
1938: Always Goodbye
1939: The Hound of the Baskervilles; Second Fiddle; Swanee River
1941: You'll Never Get Rich
1942: The Lady Has Plans; My Favorite Blonde
1943: The Meanest Man in the World; Let's Face It
1944: Standing Room Only
1945: Bring on the Girls
1946: The Well-Groomed Bride
1947: The Trouble with Women; Where There's Life
1948: Station West
1949: Sorrowful Jones
1951: The Lemon Drop Kid; Follow the Sun
1952: Skirts Ahoy!

FRITZ LANG (1890-1976) Germ.

(American films only)
1936: Fury
1937: You Only Live Once
1938: You and Me
1940: The Return of Frank James
1941: Man Hunt; Western Union; Confirm or Deny (unc. w/ A. Mayo)
1942: Moontide (w/ A. Mayo)
1943: Hangmen Also Die
1944: Ministry of Fear
1945: Scarlet Street; The Woman in the Window
1946: Cloak and Dagger
1948: Secret Beyond the Door
1950: American Guerilla in the Philippines; House by the River
1952: Clash by Night; Rancho Notorious
1953: The Big Heat; The Blue Gardenia
1954: Human Desire
1955: Moonfleet
1956: Beyond a Reasonable Doubt; While the City Sleeps
1960: Journey to the Lost City (Germ.); The Thousand Eyes of Dr. Mabuse (Germ.)

WALTER LANG (1896-1972)

- 1929: Spirit of Youth
- 1930: Hello, Sister; Cock o' the Walk (w/ Roy William Neill) The Big Fight; Brothers; The Costello Case
- 1931: Command Performance; Hell Bound; Women Go on Forever
- 1932: Meet the Baron; No More Orchids
- 1933: The Warrior's Husband
- 1934: Whom the Gods Destroy; The Party's Over; The Mighty Barnum
- 1935: Carnival; Hooray for Love
- 1936: Love Before Breakfast
- 1937: Wife, Doctor and Nurse; Second Honeymoon
- 1938: The Baroness and the Butler; I'll Give a Million
- 1939: The Little Princess
- 1940: The Blue Bird; Star Dust; The Great Profile; Tin Pan Alley
- 1941: Moon Over Miami; Week-End in Havana
- 1942: Song of the Islands; The Magnificent Dope
- 1943: Coney Island
- 1944: Greenwich Village
- 1945: State Fair
- 1946: Sentimental Journey; Claudia and David
- 1947: Mother Wore Tights
- 1948: Sitting Pretty; When My Baby Smiles at Me
- 1949: You're My Everything
- 1950: Cheaper by the Dozen; The Jackpot
- 1951: On the Riviera
- 1952: With a Song in My Heart
- 1953: Call Me Madam
- 1954: There's No Business Like Show Business
- 1956: The King and I
- 1957: Desk Set
- 1959: But Not for Me
- 1960: Can-Can; The Marriage-Go-Round
- 1961: Snow White and the Three Stooges

NOEL LANGLEY (1911-) Brit.

(American films only)
- 1956: The Search for Bridey Murphy

LARRY LANSBURGH

- 1964: The Tattooed Police Horse

ANTHONY M. LANZA

- 1967: The Glory Stompers
- 1971: The Incredible Two-Headed Transplant

CHRISTOPHER LARKIN

1974: A Very Natural Thing

JOHN LARKIN

1942: Quiet Please, Murder
1945: Circumstantial Evidence

KEITH LARSEN

1969: Mission Batangas

STAN LATHAM

1973: Save the Children (doc.); Detroit 9000 (w/ A. Marks)
1974: Amazing Grace

TOM LAUGHLIN (also T. C. FRANK) (1938-)

1965: The Young Sinner
1967: Born Losers
1972: Billy Jack
1974: The Trial of Billy Jack
1975: The Master Gunfighter
1977: Billy Jack Goes to Washington

CHARLES LAUGHTON (1899-1962)

1955: The Night of the Hunter

ARNOLD LAVEN (1922-)

1952: Without Warning
1953: Vice Squad
1954: Down Three Dark Streets
1956: The Rack
1957: The Monster That Challenged the World; Slaughter on Tenth Avenue
1958: Anna Lucasta
1962: Geronimo
1965: The Glory Guys
1967: Rough Night in Jericho
1969: Sam Whiskey

HAROLD LAW

1936: Neighborhood House

EDMUND LAWRENCE (Brit.)

(American films only)
1929: The House of Secrets

MARC LAWRENCE (1910-)

1965: Nightmare in the Sun

ASHLEY LAZARUS

1977: Golden Rendezvous

PHILIP LEACOCK (1917-) Brit.

(American films only)
1959: The Rabbit Trap
1960: Let No Man Write My Epitaph; Take a Giant Step
1961: 13 West Street
1962: The War Lover

TONY LEADER

1970: The Cockeyed Cowboys of Calico County

NORMAN LEAR (1922-)

1971: Cold Turkey

WILLIAM LE BARON

1929: The Very Idea (w/ Richard Rosson)

REGINALD LE BORG (1902-)

1943: She's For Me; Calling Dr. Death
1944: Weird Woman; The Mummy's Ghost; Jungle Woman; San Diego, I Love You; Dead Man's Eyes; Destiny (w/ Julien Duvivier); Adventure in Music (w/ others)
1945: Honeymoon Ahead

1946: Joe Palooka, Champ; Little Iodine; Susie Steps Out
1947: Fall Guy; The Adventures of Don Coyote; Philo Vance's Secret Mission; Joe Palooka in the Knockout
1948: Port Said; Joe Palooka in Winner Take All
1949: Fighting Fools; Hold That Baby!; Joe Palooka in the Counterpunch
1950: Young Daniel Boone; Wyoming Mail; Joe Palooka in the Squared Circle
1951: G.I. Jane; Joe Palooka in Triple Cross
1952: Models, Inc.
1953: Bad Blonde; The Great Jesse James Raid; Sins of Jezebel
1954: The White Orchid
1956: The Black Sleep
1957: Voodoo Island; War Drums; The Dalton Girls
1961: The Flight That Disappeared
1962: Deadly Duo
1963: Diary of a Madman
1964: The Eyes of Annie Jones (Brit.)

HERBERT J. LEDER

1960: Pretty Boy Floyd
1967: It; The Frozen Dead
1969: The Candy Man

PAUL LEDER

1974: I Disremember Mama
1976: A*P*E*

CHARLES LEDERER (c. 1906-1976)

1942: Fingers at the Window
1951: On the Loose
1959: Never Steal Anything Small

D. ROSS LEDERMAN (1895-1972)

1929: The Million Dollar Collar
1930: The Man Hunter
1931: The Texas Ranger; Branded; Range Feud
1932: Ridin' for Justice; The Fighting Marshal; High Speed; Riding Tornado; The Texas Cyclone; Daring Danger; Two-Fisted Law; McKenna of the Mounted
1933: Speed Demon; End of the Trail; Whirlwind; The State Trooper; Soldiers of the Storm; Rusty Rides Alone; Silent Men
1934: Hell Bent for Love; The Crime of Helen Stanley; A Man's Game; Beyond the Law; Girl in Danger; Murder in the Clouds
1935: Red Hot Tires; Dinky (w/ Howard Bretherton); Moonlight on the Prairie; The Case of the Missing Man; Too Tough to Kill

1936: Hell-Ship Morgan; Panic on the Air; Pride of the Marines; The Final Hour; Alibi for Murder; Come Close, Folks
1937: Counterfeit Lady; I Promise to Pay; Motor Madness; The Frame-Up; The Game That Kills
1938: Juvenile Court; The Little Adventuress; Adventure in Sahara
1939: North of Shanghai; Racketeers of the Range
1940: Military Academy; Thundering Frontier; Glamour for Sale
1941: Across the Sierras; Father's Son; Strange Alibi; Shadows on the Stairs; Passage from Hong Kong
1942: The Body Disappears; Bullet Scars; I Was Framed; Escape from Crime; Busses Roar; The Gorilla Man
1943: Adventure in Iraq; Find the Blackmailer
1944: The Racket Man
1946: The Phantom Thief; Out of the Depths; The Notorious Lone Wolf; Dangerous Business; Sing While You Dance; Boston Blackie and the Law
1947: The Lone Wolf in Mexico; Key Witness
1948: The Return of the Whistler
1950: Military Academy with That 10th Avenue Gang

(Serial)

1931: The Phantom of the West

NORMAN LEE

1933: Money Talks
1937: Bulldog Drummond at Bay
1938: Kathleen

ROWLAND V. LEE (1891-1975)

1929: Wolf of Wall Street; A Dangerous Woman; Mysterious Dr. Fu Manchu
1930: Paramount on Parade (w/ others); The Return of Dr. Fu Manchu; Ladies Love Brutes; A Man from Wyoming; Derelict
1931: The Ruling Voice; The Guilty Generation; Upper Underworld
1933: Zoo in Budapest
1934: I Am Suzanne; The Count of Monte Cristo; Over Night (Brit.: That Night in London); Gambling
1935: Cardinal Richelieu; The Three Musketeers
1936: One Rainy Afternoon
1937: Love from a Stranger (Brit.); The Toast of New York
1938: Mother Carey's Chickens; Service De Luxe
1939: Son of Frankenstein; The Sun Never Sets; Tower of London
1940: The Son of Monte Cristo
1942: Powder Town
1944: The Bridge of San Luis Rey
1945: Captain Kidd

HERBERT I. LEEDS (-1954)

1938: Love on a Budget; Island in the Sky; Keep Smiling; Five of a Kind; Arizona Wildcat
1939: Mr. Moto in Danger Island; The Return of the Cisco Kid; Chicken Wagon Family; Charlie Chan in City in Darkness
1940: Cisco Kid and the Lady; Yesterday's Heroes
1941: Romance of the Rio Grande; Ride on Vaquero; Blue, White and Perfect
1942: The Man Who Wouldn't Die; Just off Broadway; Manila Calling; Time to Kill
1946: It Shouldn't Happen to a Dog
1948: Let's Live Again
1950: Bunco Squad; Father's Wild Game

JACK LEEWOOD

1961: 20,000 Eyes
1963: Thunder Island

ED LEFTWICH

1960: Squad Car

ERNEST LEHMAN (1920-)

1972: Portnoy's Complaint

HENRY LEHRMAN (1886-1946)

1929: Homesick

MAX LEIBMAN

1973: Ten from Your Show of Shows

MITCHELL LEISEN (1898-1972)

1933: Cradle Song
1934: Death Takes a Holiday; Murder at the Vanities
1935: Behold My Wife; Hands Across the Table; Four Hours to Kill
1936: The Big Broadcast of 1937; 13 Hours by Air
1937: Easy Living; Swing High, Swing Low
1938: Artists and Models Abroad; The Big Broadcast of 1938
1939: Midnight

1940: Arise, My Love; Remember the Night
1941: Hold Back the Dawn; I Wanted Wings
1942: The Lady Is Willing; Take a Letter, Darling
1943: No Time for Love
1944: Frenchman's Creek; Lady in the Dark; Practically Yours
1945: Masquerade in Mexico; Kitty
1946: To Each His Own
1947: Golden Earrings; Suddenly, It's Spring
1948: Dream Girl
1949: Bride of Vengeance; Song of Surrender
1950: Captain Carey, USA; No Man of Her Own
1951: Darling, How Could You!; The Mating Season
1952: Young Man with Ideas
1953: Tonight We Sing
1955: Bedevilled
1957: The Girl Most Likely
1967: Spree! (doc. w/ W. Green)

CHRISTOPHER LEITCH

1979: The Hitter

ALAN LEMAY (1899-)

1950: High Lonesome

JACK LEMMON (1925-)

1971: Kotch

JOHN LEMONT

1955: The Green Buddha
1960: The Shakedown

PAUL LENI (1885-1929)

(American films only)
1929: The Last Warning

ARTHUR LEONARD

1939: Straight to Heaven
1940: Pocomania
1947: Boy! It's a Girl; Sepia Cinderella

HERBERT LEONARD

1967: The Perils of Pauline (w/ J. Shelley)
1971: Going Home

ROBERT Z. LEONARD (1889-1968)

1929: Marianne
1930: The Divorcee; In Gay Madrid; Let Us Be Gay
1931: The Bachelor Father; It's a Wise Child; Five and Ten; Susan Lennox, Her Rise and Fall
1932: Lovers Courageous; Strange Interlude
1933: Peg o' My Heart; Dancing Lady
1935: After Office Hours; Escapade
1936: The Great Ziegfeld; Piccadilly Jim
1937: Maytime; The Firefly
1938: The Girl of the Golden West
1939: Broadway Serenade
1940: New Moon; Pride and Prejudice; Third Finger, Left Hand
1941: Ziegfeld Girl; When Ladies Meet
1942: We Were Dancing; Stand by for Action
1943: The Man from Down Under
1944: Marriage Is a Private Affair
1945: Weekend at the Waldorf
1946: The Secret Heart
1947: Cynthia
1948: B. F.'s Daughter
1949: The Bribe; In the Good Old Summertime
1950: Nancy Goes to Rio; Duchess of Idaho; Grounds for Marriage
1951: Too Young to Kiss
1952: Everything I Have Is Yours
1953: The Clown; The Great Diamond Robbery
1954: Her Twelve Men
1955: The King's Thief
1956: Kelly and Me
1958: Beautiful but Dangerous (Ital.)

JORDAN LEONDOPOULOS

1969: Sam's Song

JOHN LEONE

1976: The Great Smokey Roadblock

CARL LERNER

1964: Black Like Me

IRVING LERNER (1909-1976)

1951: Suicide Attack (doc.)
1953: Man Crazy
1958: Murder by Contract; Edge of Fury
1959: City of Fear
1960: Studs Lonigan
1963: Cry of Battle
1969: Royal Hunt of the Sun (Brit.)

JOSEPH LERNER

1947: The Fight Never Ends
1949: C-Man
1950: Guilty Bystander; Mr. Universe

MURRAY LERNER

1956: Secrets of the Reef (doc.)
1967: Festival (doc.)

RICHARD LERNER

1976: Revenge of the Cheerleaders

MERVYN LEROY (1900-)

1929: Broadway Babies; Hot Stuff; Naughty Baby
1930: Little Caesar; Playing Around; Numbered Men; Show Girl in Hollywood; Top Speed; Little Johnny Jones
1931: Broad-Minded; Five Star Final; Local Boy Makes Good; Too Young to Marry; Gentleman's Fate; Tonight or Never
1932: Big City Blues; I Am a Fugitive from a Chain Gang; Heart of New York; High Pressure; 3 on a Match; Two Seconds
1933: Elmer the Great; Gold Diggers of 1933; Hard to Handle; The World Changes; Tugboat Annie
1934: Hi, Nellie; Heat Lightning; Happiness Ahead; Sweet Adeline
1935: I Found Stella Parish; Oil for the Lamps of China; Page Miss Glory
1936: Anthony Adverse; Three Men on a Horse
1937: The King and the Chorus Girl; They Won't Forget
1938: Fools for Scandal
1940: Escape; Waterloo Bridge
1941: Blossoms in the Dust; Johnny Eager; Unholy Partners
1942: Random Harvest
1943: Madame Curie

1944: 30 Seconds over Tokyo
1946: Without Reservations
1948: Homecoming
1949: Any Number Can Play; East Side, West Side; Little Women
1951: Quo Vadis
1952: Million Dollar Mermaid; Lovely to Look At
1953: Latin Lovers
1954: Rose Marie
1955: Strange Lady in Town
1956: The Bad Seed; Toward the Unknown
1958: Home Before Dark; No Time for Sergeants
1959: The FBI Story
1960: Wake Me When It's Over
1961: The Devil at 4 O'Clock; A Majority of One
1962: Gypsy
1963: Mary, Mary
1966: Moment to Moment
1969: Downstairs at Ramsey's
1970: The 13 Clocks

MARK L. LESTER

1973: Steel Arena
1974: Truck Stop Women
1976: Bobbie Jo and the Outlaw
1977: Stunts (ret. Who Is Killing the Stuntmen?)
1979: Roller Boogie

RICHARD LESTER (1932-   ) Brit.

(American films only)
1966: A Funny Thing Happened on the Way to the Forum
1968: Petulia
1976: The Ritz
1979: Butch & Sundance: The Early Years; Cuba

DICK L'ESTRANGE

1944: Teen Age

JOSEPH LEVERING

1931: Sea Devils; Defenders of the Law
1933: Cheating Blondes
1938: Frontiers of '49; In Early Arizona; Pioneer Trail; Phantom Gold; Rolling Caravans; Stagecoach Days
1939: The Law Comes to Texas; Lone Star Pioneers

MICHEL LEVESQUE

1971: Werewolves on Wheels
1973: Sweet Sugar

WILLIAM A. LEVEY

1977: The Happy Hooker Goes to Washington
1979: Skatetown, U.S.A.

HENRY LEVIN (1909-)

1944: Cry of the Werewolf; Sergeant Mike
1945: Dancing in Manhattan; I Love a Mystery
1946: The Fighting Guardsman; The Bandit of Sherwood Forest (w/ George Sherman); Night Editor; The Unknown; The Devil's Mask; The Return of Monte Cristo
1947: The Guilt of Janet Ames; The Corpse Came C.O.D.
1948: The Mating of Millie; The Gallant Blade; The Man from Colorado
1949: Mr. Soft Touch (w/ Gordon Douglas); Jolson Sings Again
1950: And Baby Makes Three; Convicted; The Petty Girl; The Flying Missile
1951: Two of a Kind; The Family Secret
1952: Belles on Their Toes
1953: The President's Lady; The Farmer Takes a Wife; Mister Scoutmaster
1954: Three Young Texans; The Gambler from Natchez
1955: The Warriors
1957: The Lonely Man; Let's Be Happy (Brit.); Bernardine; April Love
1958: A Nice Little Bank That Should Be Robbed
1959: The Remarkable Mr. Pennypacker; Holiday for Lovers; Journey to the Center of the Earth
1960: Where the Boys Are
1961: The Wonders of Aladdin (Ital.)
1962: The Wonderful World of the Brothers Grimm (w/ George Pal; Levin directed narrative sequences); If a Man Answers
1963: Come Fly with Me
1964: Honeymoon Hotel
1965: Genghis Khan
1966: Murderers' Row
1967: Kiss the Girls and Make Them Die (Ital.); The Ambushers
1969: The Desperados
1973: That Man Bolt (w/ David Lowell Rich; U.S./ Hong Kong)
1978: Run for the Roses

JACK LEVIN (also JACK JEVNE)

1935: The Ghost Rider

MEYER LEVIN

1948: The Illegals (doc.)

SID LEVIN

1973: Let the Good Times Roll (w/ Bob Abel)
1978: The Great Brain

FRED LEVINSON

1973: Hail to the Chief
1979: Washington, B.C.

ABE LEVITOW

1963: Gay Purr-ee (anim.)
1970: The Phantom Tollbooth (w/ others)

RALPH LEVY

1964: Bedtime Story
1965: Do Not Disturb

ALBERT LEWIN (1894-1968)

1942: The Moon and Sixpence
1945: The Picture of Dorian Gray
1947: The Private Affairs of Bel Ami
1951: Pandora and the Flying Dutchman
1953: Saadia
1957: The Living Idol

ROBERT LEWIN

1961: Third of a Man

AL LEWIS

1956: Our Miss Brooks

EDGAR LEWIS

1929: Unmasked
1930: Ladies in Love; Love at First Sight

HERSCHELL GORDON LEWIS (1926-)

1961: Living Venus; Lucky Pierre
1962: Nature's Playmates; Daughters of the Sun; B-O-I-N-N-G!
1963: Scum of the Earth; Goldilocks and the Three Bares; Blood Feast
1964: Bell, Bare, and Beautiful; Two Thousand Maniacs!; Color Me Blood Red
1965: Monster-A-Go-Go!; Alley Tramp; Moonshine Mountain
1966: Jimmy, the Boy Wonder; Sin, Suffer and Repent; The Magic Land of Mother Goose
1967: Suburban Roulette; The Pill; The Gruesome Twosome; Blast-off Girls; How to Make a Doll; A Taste of Blood
1968: Something Weird; Just for the Hell of It; She-Devils on Wheels
1969: The Ecstasies of Women; Miss Nymphet's Zap-In
1970: The Wizard of Gore
1971: This Stuff'll Kill Ya'!
1972: Year of the Yahoo; The Gore-Gore Girls (Blood Orgy); Stick It in Your Ear

JERRY LEWIS (1926-)

1960: The Bellboy
1961: The Ladies' Man
1962: The Errand Boy
1963: The Nutty Professor
1964: The Patsy
1965: The Family Jewels
1966: Three on a Couch
1967: The Big Mouth
1970: One More Time; Which Way to the Front?

JOSEPH H. LEWIS (1900-)

1937: Navy Spy (w/ Crane Wilbur); The Singing Outlaw; Courage of the West
1938: The Spy Ring (International Spy); Border Wolves; The Last Stand
1940: Two-Fisted Rangers; Blazing Six Shooters; Texas Stage-Coach; The Man from Tumbleweeds; The Return of Wild Bill; Boys of the City; That Gang of Mine
1941: Pride of the Bowery; The Invisible Ghost; Criminals Within; Arizona Cyclone
1942: Bombs over Burma; The Silver Bullet; Secrets of a Co-ed (Silent Witness); The Boss of Hangtown Mesa; The Mad Doctor of Market Street
1944: Minstrel Man
1945: The Falcon in San Francisco; My Name Is Julia Ross
1946: So Dark the Night; The Jolson Story (directed musical numbers)

1947: The Swordsman
1948: The Return of October
1949: The Undercover Man; Gun Crazy (Deadly Is the Female)
1950: A Lady Without Passport
1952: Retreat, Hell!; Desperate Search
1953: Cry of the Hunted
1955: The Big Combo; A Lawless Street
1956: The Seventh Cavalry
1957: The Halliday Brand
1958: Terror in a Texas Town

ROBERT LEWIS

1956: Anything Goes

VANCE LEWIS

1975: The Silent Stranger

JOSEF LEYTES

1967: Valley of Mystery
1968: The Counterfeit Killer

ART LIEBERMAN

1975: Up Your Alley
1979: The Melon Affair

JEFF LIEBERMAN

1976: Squirm; Blue Sunshine

EDWARD C. LILLEY (1896-1974)

1942: Cross Your Fingers; Never a Dull Moment
1943: Honeymoon Lodge; Larceny with Music; Moonlight in Vermont
1944: Allergic to Love; Babes on Swing Street; Hi, Good Lookin'; My Gal Loves Music; Sing a Jingle
1945: Her Lucky Night; Swing Out, Sister

JOHN F. LINK

1947: Call of the Forest
1948: Devil's Cargo

ROBERT L. LIPPERT

1948: Last of the Wild Horses

ANATOLE LITVAK (MICHAEL ANATOL LITWAK) (1902-1974)

1931: Dolly Gets Ahead (Germ.)
1933: Be Mine Tonight (Germ.); Sleeping Car (Brit.)
1937: The Woman I Love; Mayerling (Fr.); Tovarich
1938: The Amazing Dr. Clitterhouse; The Sisters; Flight into Darkness (Fr.)
1939: Castle on the Hudson; Confessions of a Nazi Spy
1940: All This and Heaven Too; City for Conquest
1941: Out of the Fog; Blues in the Night
1942: This Above All
1947: The Long Night
1948: Sorry, Wrong Number; The Snake Pit
1952: Decision Before Dawn
1953: Act of Love (Fr.)
1955: The Deep Blue Sea (Brit.)
1956: Anastasia
1959: The Journey (Austria)
1961: Goodbye Again (Fr.)
1963: Five Miles to Midnight (Fr.)
1967: The Night of the Generals (Germ.)
1970: The Lady in the Car with Glasses and a Gun (Fr.)

FRANK LLOYD (1888-1960)

1929: Weary River; The Divine Lady; Drag; Young Nowheres; Dark Streets
1930: Son of the Gods; The Way of All Men
1931: The Lash; East Lynne; The Right of Way; The Age for Love
1932: A Passport to Hell
1933: Cavalcade; Berkeley Square; Hoopla
1934: Servants' Entrance
1935: Mutiny on the Bounty
1936: Under Two Flags
1937: Maid of Salem; Wells Fargo
1938: If I Were King
1939: Rulers of the Sea
1940: The Howards of Virginia
1941: The Lady from Cheyenne; This Woman Is Mine
1943: Forever and a Day (w/ others)
1945: Blood on the Sun
1954: The Shanghai Story
1955: The Last Command

PETER LOCKE

1971: You've Got to Walk It Like You Talk It or You'll Lose That Beat

ROY LOCKWOOD

1957: Jamboree

BARBARA LODEN

1971: Wanda

ANTHONY LOEB

1974: As Time Goes By (doc.)

JOSHUA LOGAN (1908-)

1938: I Met My Love Again (w/ Arthur Ripley)
1955: Picnic
1956: Bus Stop
1957: Sayonara
1958: South Pacific
1960: Tall Story
1961: Fanny
1964: Ensign Pulver
1967: Camelot
1969: Paint Your Wagon

STANLEY LOGAN

1937: First Lady
1938: Love, Honor and Behave; Women Are Like That
1942: The Falcon's Brother

LOU LOMBARDO

1975: Russian Roulette

ULLI LOMMEL

1979: Cocaine Cowboys

DEL LORD

1929: Barnum Was Right
1936: Trapped by Television
1937: What Price Vengeance

1944: Kansas City Kitty; She's a Sweetheart
1945: I Love a Bandleader; Rough, Tough, and Ready; Blonde from Brooklyn; Hit the Hay
1946: In Fast Company; It's Great to Be Young; Singin' in the Corn

PARE LORENTZ (1905-)

1936: The Plow That Broke the Plains (doc.)
1937: The River (doc.)
1940: The Fight for Life (doc.)

ANTON LORENZ

1934: Back Page

THOMAS Z. LORING

1942: Who Is Hope Schuyler?; Thru Different Eyes
1943: He Hired the Boss

JOSEPH LOSEY (1909-)

1948: The Boy with Green Hair
1950: The Lawless
1951: The Big Night; M; The Prowler
1952: Stranger on the Prowl (Ital.)
1954: The Sleeping Tiger (Brit.)
1955: Finger of Guilt (Brit.)
1956: Town Without Pity (Brit.)
1957: The Gypsy and the Gentleman (Brit.)
1958: Chance Meeting (Brit.)
1961: These Are the Damned (Brit.)
1962: The Concrete Jungle (Brit.); Eve (Ital.)
1963: The Servant (Brit.)
1964: King and Country (Brit.)
1966: Modesty Blaise (Brit.)
1967: Accident (Brit.)
1968: Boom; Sweet Ceremony (Brit.)
1970: Figures in a Landscape
1971: The Go-Between (Brit.)
1972: The Assassination of Trotsky
1973: A Doll's House
1975: The Romantic Englishwoman (Brit.); Galileo
1977: Mr. Klein (Fr.)
1978: The Roads to the South
1979: Don Giovanni

EUGENE LOURIE (1905-)

1953: The Beast from 20,000 Fathoms; The Colossus of New York
1959: The Giant Behemoth (w/ Douglas Hickox; Brit.)
1961: Gorgo (Brit.)

ANTHONY LOVER

1975: Distance

OTHO LOVERING

1934: Wanderer of the Wasteland
1936: The Sky Parade; Border Flight; Drift Fence

ARTHUR LUBIN (1901-)

1934: A Successful Failure
1935: Great God Gold; Honeymoon Limited; Two Sinners; Frisco Waterfront
1936: The House of a Thousand Candles; Yellowstone
1937: Mysterious Crossing; California Straight Ahead; I Cover the War; Idol of the Crowds; Adventure's End
1938: Midnight Intruder; Beloved Brat; Prison Break; Secrets of a Nurse
1939: Risky Business; Big Town Czar; Mickey the Kid; Call a Messenger
1940: The Big Guy; Black Friday; Gangs of Chicago; I'm Nobody's Sweetheart Now; Meet the Wildcat; Who Killed Aunt Maggie?
1941: San Francisco Docks; Where Did You Get That Girl?; Buck Privates; In the Navy; Hold That Ghost; Keep 'Em Flying
1942: Ride 'Em Cowboy; Eagle Squadron
1943: White Savage; The Phantom of the Opera
1944: Ali Baba and the Forty Thieves
1945: Delightfully Dangerous
1946: The Spider Woman Strikes Back; A Night in Paradise
1947: New Orleans
1949: Impact; Francis
1951: Queen for a Day; Francis Goes to the Races; Rhubarb
1952: Francis Goes to West Point; It Grows on Trees
1953: South Sea Woman; Francis Covers the Big Town
1954: Francis Joins the WACS
1955: Francis in the Navy; Footsteps in the Fog; Lady Godiva
1956: Star of India; The First Traveling Saleslady
1957: Escapade in Japan
1961: Thief of Baghdad
1964: The Incredible Mr. Limpet
1966: Hold On!
1971: Rain for a Dusty Summer (Span.)

ERNST LUBITSCH (1892-1947)

(American films only)
1929: The Love Parade; Eternal Love
1930: Monte Carlo; Paramount on Parade (w/ others)
1931: The Smiling Lieutenant
1932: Broken Lullaby (ret. The Man I Killed); One Hour with You (w/ G. Cukor); Trouble in Paradise; If I Had a Million (w/ others)
1933: Design for Living
1934: The Merry Widow
1937: Angel
1938: Bluebeard's Eighth Wife
1939: Ninotchka
1940: The Shop Around the Corner
1941: That Uncertain Feeling
1942: To Be or Not to Be
1943: Heaven Can Wait
1945: A Royal Scandal (w/ O. Preminger)
1946: Cluny Brown
1948: That Lady in Ermine (w/ O. Preminger, unc.)

S. ROY LUBY

1935: Outlaw Rule; Range Warfare; Lightning Triggers
1936: The Crooked Trail; The Desert Phantom; Rogue of the Range
1937: Border Phantom; Race Suicide; The Red Rope; Tough to Handle
1940: The Range Busters; Trailing Double Trouble; West of Pinto Basin
1941: Fugitive Valley; The Kid's Last Ride; Trail of the Silver Spurs; Tumbledown Ranch in Arizona; Wrangler's Roost; Saddle Mountain Roundup; Tonto Basin Outlaws; Underground Rustlers
1942: Pride of the Army (orig. War Dogs); Boot Hill Bandits; Rock River Renegades; Texas Trouble Shooters; Thunder River Feud; Arizona Stagecoach
1943: Black Market Rustlers; Cowboy Commandos; Land of Hunted Men

GEORGE LUCAS (1945-)

1971: THX-1138
1973: American Graffiti
1977: Star Wars

EDWARD LUDWIG (c.1900-)

1932: Steady Company

1933: They Just Had to Get Married
1934: A Woman's Man; Let's Be Ritzy; Friends of Mr. Sweeney
1935: The Man Who Reclaimed His Head; Age of Indiscretion; Old Man Rhythm; Three Kids and a Queen
1936: Fatal Lady; Adventure in Manhattan
1937: Her Husband Lies; The Last Gangster
1938: That Certain Age
1939: Coast Guard
1940: Swiss Family Robinson
1941: The Man Who Lost Himself
1942: Born to Sing
1943: They Came to Blow Up America
1944: The Fighting Seabees; Three Is a Family
1947: The Fabulous Texan
1948: Wake of the Red Witch
1949: The Big Wheel
1951: Smuggler's Island
1952: Caribbean; Big Jim McLain; The Blazing Forest
1953: The Vanquished; Sangaree
1954: Jivaro
1955: Flame of the Islands
1957: The Black Scorpion
1963: The Gun Hawk

SIDNEY LUMET (1924-   )

1956: Twelve Angry Men
1958: Stage Struck
1959: The Fugitive Kind; That Kind of Woman
1961: A View from the Bridge
1962: A Long Day's Journey into Night
1963: Fail-Safe
1965: The Hill; The Pawnbroker
1966: The Group
1967: The Deadly Affair
1968: Bye Bye Braverman; The Sea Gull (Brit.)
1969: The Appointment
1970: King: A Filmed Record ... Montgomery to Memphis (doc. w/ J. L. Mankiewicz); Last of the Mobile Hot-Shots
1972: The Anderson Tapes; Child's Play
1973: The Offense (Brit.); Serpico
1974: Lovin' Molly; Murder on the Orient Express (Brit.)
1975: Dog Day Afternoon
1976: Network
1977: Equus
1978: The Wiz
1979: Just Tell Me What You Want

IDA LUPINO (1916-   )

1950: Never Fear; Outrage

1951: Hard, Fast and Beautiful; On Dangerous Ground (unc. w/ Nicholas Ray)
1953: The Hitch-Hiker; The Bigamist
1966: The Trouble with Angels

HAMILTON LUSKE

1941: The Reluctant Dragon (anim. w/ others)
1951: Alice in Wonderland (anim. w/ others)
1953: Peter Pan (anim. w/ others)
1955: Lady and the Tramp (anim. w/ others)

SIDNEY B. LUST

1934: The Birth of a New America (doc.)

HENRY LYNN

1939: Mothers of Today

BURT LYNWOOD

1935: Motive for Revenge; The Firetrap; Reckless Roads
1937: Shadows of the Orient

FRANCIS D. LYON (1905-)

1953: Crazylegs
1954: The Bob Mathias Story
1955: The Cult of the Cobra
1956: The Great Locomotive Chase
1957: The Oklahoman; Bailout at 43,000; Gunsight Ridge
1958: South Seas Adventure (w/ others)
1959: Escort West
1961: Tomboy and the Champ
1963: The Young and the Brave
1966: Destination Inner Space
1967: Castle of Evil
1968: The Destructors; The Money Jungle
1969: The Girl Who Knew Too Much

BERT LYTELL

1936: Along Came Love (w/ D. Mansfield)

CHARLES MACARTHUR (1895-1950)

1934: Crime Without Passion (w/ Ben Hecht)
1935: The Scoundrel (w/ B. Hecht)
1936: Soak the Rich (w/ B. Hecht); Once in a Blue Moon (w/ B. Hecht)

JIM McBRIDE

1974: Hot Times

GENE McCABE

1959: Follow Me

ROBERT McCAHON

1973: Running Wild
1975: Deliver Us from Evil

LEO McCAREY (1898-1969)

1929: The Sophomore; Red Hot Rhythm
1930: Wild Company; Let's Go Native; The Sheeper-Newfounder (alt. Part-Time Wife)
1931: Indiscreet
1932: The Kid from Spain
1933: Duck Soup
1934: Belle of the Nineties; Six of a Kind
1935: Ruggles of Red Gap
1936: The Milky Way
1937: The Awful Truth; Make Way for Tomorrow
1939: Love Affair
1942: Once Upon a Honeymoon
1944: Going My Way
1945: The Bells of St. Mary's
1948: Good Sam
1952: My Son John
1957: An Affair to Remember
1958: Rally 'Round the Flag, Boys!
1962: Satan Never Sleeps

RAY McCAREY (1904-1948)

1932: Pack Up Your Troubles (w/ George Marshall)
1934: Girl of My Dreams
1935: Millions in the Air; Sunset Range; Mystery Man; Hot Tip (w/ James Gleason)

1936: Three Cheers for Love
1937: Oh, Doctor; Let's Make a Million; Love in a Bungalow
1938: Goodbye, Broadway; The Devil's Party
1939: Torchy Runs for Mayor
1940: You Can't Fool Your Wife; Millionaires in Prison; Little Orvie
1941: Accent on Love; Cadet Girl; The Cowboy and the Blonde; The Perfect Snob; Murder Among Friends
1942: A Gentleman at Heart; It Happened in Flatbush; That Other Woman
1943: So This Is Washington
1944: Passport to Adventure (orig. Passport to Destiny); Atlantic City
1946: The Falcon's Alibi; Strange Triangle
1948: The Gay Intruders

JOHN P. McCARTHY (1885-)

1931: Cavalier of the West; Nevada Buckaroo; Rose of the Rio Grande; The Ridin' Fool; Rider of the Plains; Sunrise Trail; God's Country and the Man; Ships of Hate; Mother and Son
1932: The Western Code; The Fighting Champ; The '49ers
1933: Trailin' North; Lucky Larrigan; Return of Casey Jones
1935: The Lawless Border
1936: Song of the Gringo
1944: Marked Trails; Raiders of the Border
1945: The Cisco Kid Returns

ROBERT McCARTY

1973: I Could Never Have Sex with Any Man Who Has So Little Respect for My Husband

JOHN McCAULEY

1976: Rattlers

GUTHRIE McCLINTIC

1930: On Your Back
1931: Once a Sinner; Once a Lady

GEORGE McCOWAN

1972: Frogs; The Magnificent Seven Ride!
1976: Shadow of the Hawk (Can.)
1979: The Shape of Things to Come (Can.)

DENYS McCOY

1971: The Last Rebel

HANK McCUNE

1956: Wetbacks

FRANK McDONALD (1899-)

1935: Broadway Hostess
1936: The Murder of Dr. Harrigan; Boulder Dam; The Big Noise; Love Begins at Twenty; Treachery Rides the Range; Murder by an Aristocrat; Smart Blonde; Isle of Fury
1937: Midnight Court; Her Husband's Secretary; Fly-Away Baby; Dance, Charlie, Dance; The Adventurous Blonde
1938: Blondes at Work; Reckless Living; Over the Wall; Freshman Year; Flirting with Fate
1939: First Offenders; They Asked for It; Jeepers Creepers
1940: Rancho Grande; In Old Missouri; Gaucho Serenade; Carolina Moon; Ride, Tenderfoot, Ride; Grand Ole Opry; Barnyard Follies
1941: Arkansas Judge; Country Fair; Flying Blind; Under Fiesta Stars; Tuxedo Junction; No Hands on the Clock
1942: Shepherd of the Ozarks; The Old Homestead; Mountain Rhythm; Wildcat; Wrecking Crew; The Traitor Within
1943: High Explosive; Swing Your Partner; Alaska Highway; Submarine Alert; Hoosier Holiday; O, My Darling Clementine
1944: Timber Queen; Take It Big; Sing Neighbor, Sing; One Body Too Many; Lights of Old Santa Fe
1945: Bells of Rosarita; The Chicago Kid; The Man from Oklahoma; Tell It to a Star; Sunset in El Dorado; Along the Navajo Trail
1946: Song of Arizona; Rainbow over Texas; My Pal Trigger; Sioux City Sue
1947: Hit Parade of 1947; Twilight on the Rio Grande; Under Nevada Skies; Bulldog Drummond Strikes Back; When a Girl's Beautiful; Linda Be Good
1948: Mr. Reckless; 13 Lead Soldiers; French Leave; Gun Smugglers
1949: The Big Sombrero; Ringside; Apache Chief
1950: Snow Dog; Call of the Klondike
1951: Sierra Passage; Texans Never Cry; Father Takes the Air; Yukon Manhunt; Yellow Fin; Northwest Territory
1952: Sea Tiger; Yukon Gold
1953: Son of Belle Starr; Border City Rustlers
1954: Thunder Pass
1955: The Treasure of Ruby Hills; The Big Tip-Off
1960: The Purple Gang; Raymie
1962: The Underwater City
1963: Gunfight at Comanche Creek
1965: Mara of the Wilderness

RANALD MacDOUGALL (1915-1973)

1955: Queen Bee
1956: Hot Cars
1957: Man on Fire
1959: The World, the Flesh and the Devil
1960: The Subterraneans
1961: Go Naked in the World

RODDY McDOWALL (1928-)

1972: The Devil's Widow (orig. Tam-Lin; Brit.)

BERNARD McEVEETY

1966: Ride Beyond Vengeance
1971: The Brotherhood of Satan
1972: Napoleon and Samantha
1973: One Little Indian
1974: The Bears and I

VINCENT McEVEETY

1968: Firecreek
1971: The Million Dollar Duck
1972: The Biscuit Eater
1973: Charley and the Angel
1974: Superdad; The Castaway Cowboy
1975: The Strangest Man in the World
1976: Gus; Treasure of Matecumbe
1977: Herbie Goes to Monte Carlo
1979: The Apple Dumpling Gang Rides Again

EARL McEVOY

1950: Cargo to Capetown; The Killer That Stalked New York
1951: The Barefoot Mailman

HAMILTON MacFADDEN (1901-)

1930: Harmony at Home; Crazy That Way; Oh, for a Man!; Are You There?
1931: Charlie Chan Carries On; The Black Camel; Riders of the Purple Sage; Their Mad Moment (w/ Chandler Sprague)
1932: Cheaters at Play
1933: Second Hand Wife; The Fourth Horseman; Trick for Trick; The Man Who Dared; Charlie Chan's Greatest Case

1934: As Husbands Go; Hold That Girl; Stand Up and Cheer; She Was a Lady
1935: Elinor Norton; Fighting Youth
1937: The Three Legionnaires; It Can't Last Forever; Sea Racketeers; Escape by Night
1942: Inside the Law

WILLIAM F. McGAHA

1972: J. C.

WILLIAM H. McGANN (1895-1977)

1930: On the Border
1931: I Like Your Nerve
1932: Illegal (Brit.); Murder on the Second Floor (Brit.)
1935: Maybe It's Love; A Night at the Ritz; Man of Iron
1936: Freshman Love; Brides Are Like That; Times Square Playboy; Two Against the World; Hot Money; Polo Joe; The Case of the Black Cat
1937: Penrod and Sam; Marry the Girl; Sh! The Octopus
1938: Alcatraz Island; Penrod and His Twin Brother; When Were You Born?; Girls on Probation
1939: Blackwell's Island; Sweepstakes Winner; Everybody's Hobby; Pride of the Blue Grass
1940: Wolf of New York; Dr. Christian Meets the Women
1941: A Shot in the Dark; The Parson of Panamint; Highway West; We Go Fast
1942: In Old California; Tombstone, the Town Too Tough to Die; American Empire
1943: Frontier Badman

W. F. McGAUGH

(Serial)

1935: The New Adventures of Tarzan (w/ Edward Kull)

DARREN McGAVIN (1922-)

1973: Happy Mother's Day, Love George (ret. Run, Stranger, Run)

PATRICK McGOOHAN (1928-)

1947: Catch My Soul

DORRELL McGOWAN

1950: The Showdown (w/ Stuart McGowan)
1951: Tokyo File 212 (w/ S. McGowan)
1958: Snowfire (w/ S. McGowan)
1962: The Bashful Elephant (w/ S. McGowan)

J. P. McGOWAN (1880-1952)

1929: The Cowboy and the Outlaw; The Invaders; Riders of the Rio Grande
1930: Breezy Bill; Call of the Desert; The Canyon of Missing Men; The Oklahoma Sheriff; Beyond the Law; Code of Honor; Covered Wagon Trails; Hunted Men; The Man from Nowhere; Near the Rainbow's End; O'Malley Rides Alone; The Parting of the Trails; Pioneers of the West; Western Honor; 'Neath Western Skies
1931: Shotgun Pass; Under Texas Skies; Riders of the North; Headin' for Trouble; Quick Trigger Lee; Cyclone Kid
1932: Human Target; Mark of the Spur; Tangled Fortunes; Scarlet Brand; Man from New Mexico
1933: Drum Taps; When a Man Rides Alone; Deadwood Pass; War of the Range
1936: The Outlaw Tamer
1937: Rough Riding Rhythm
1938: Roaring Six Guns

(Serial)

1932: The Hurricane Express (w/ A. Schaefer)

ROBERT McGOWAN

1935: Frontier Justice
1936: Too Many Parents
1940: The Haunted House; The Old Swimmin' Hole; Tomboy

STUART McGOWAN

1950: The Showdown (w/ Dorrell McGowan)
1951: Tokyo File 212 (w/ D. McGowan)
1958: Snowfire (w/ D. McGowan)
1962: The Bashful Elephant (w/ D. McGowan)
1968: The Billion Dollar Hobo

TOM McGOWAN

1956: The Amazon Trader

1958: Manhunt in the Jungle
1960: The Hound That Thought He Was a Raccoon

CHARLES McGRATH

1930: Her Unborn Child (w/ Al Ray)

EDGAR J. McGREGOR

1930: Good News (w/ N. Grinde)

SEAN McGREGOR

1974: People Toys

THOMAS McGUANE

1975: 92 in the Shade

DENNIS McGUIRE

1974: Shoot It: Black, Shoot It: Blue

DON McGUIRE (1919-)

1955: Break to Freedom
1956: Johnny Concho
1957: The Delicate Delinquent; Hear Me Good

GUSTAV MACHATY (1901-1963) Czech.

1939: Within the Law
1945: Jealousy

BRICE MACK

1978: Jennifer
1979: Half a House; Swap Meet

RAY MACK

1942: Hillbilly Blitzkrieg

RUSSELL MACK (1892-   )

1930: Second Wife; Big Money; Night Work
1931: Lonely Wives; The Spirit of Notre Dame; Heaven on Earth
1932: Once in a Lifetime; The All American; Scandal for Sale
1933: Private Jones
1934: The Band Plays On; Meanest Girl in Town

WILLARD MACK (-1933)

1929: The Voice of the City
1933: What Price Innocence?; Broadway to Hollywood

ALEXANDER MacKENDRICK (1912-   ) Brit.

(American films only)
1957: Sweet Smell of Success
1967: Don't Make Waves

KENNETH MacKENNA (1899-1962)

1931: Always Goodbye (w/ W. C. Menzies); The Spider (w/ Menzies); Good Sport
1932: Careless Lady
1933: Walls of Gold
1934: Sleepers East

KENT MacKENZIE

1971: Saturday Morning (doc.)

ANDREW V. McLAGLEN (1920-   )

1956: Gun the Man Down; The Man in the Vault
1957: The Abductors
1960: Freckles
1961: The Little Shepherd of Kingdom Come
1963: McLintock!
1965: Shenandoah
1966: The Rare Breed
1967: Monkeys, Go Home!; The Way West
1968: The Ballad of Josie; Bandolero!; The Devil's Brigade
1969: Hellfighters; The Undefeated
1970: Chisum
1971: One More Train to Rob; Fools' Parade; Something Big
1973: Cahill: U. S. Marshal
1975: Mitchell
1976: The Last Hard Man

1978: The Wild Geese
1979: North Sea Highjack

SHIRLEY MacLAINE (1934-   )

1975: The Other Half of the Sky: A China Memoir (doc. w/ Claudia Weill)

NORMAN Z. McLEOD (1898-1964)

1931: Along Came Youth (w/ Lloyd Corrigan); Finn and Hattie (w/ Norman Taurog); Monkey Business; Touchdown
1932: The Miracle Man; Horse Feathers; If I Had a Million (w/ others)
1933: A Lady's Profession; Mama Loves Papa; Alice in Wonderland
1934: Melody in Spring; Many Happy Returns; It's a Gift
1935: Redheads on Parade; Here Comes Cookie; Coronado
1936: Early to Bed; Pennies from Heaven; Mind Your Own Business
1937: Topper
1938: Merrily We Live; There Goes My Heart
1939: Topper Takes a Trip; Remember?
1940: Little Men
1941: The Trial of Mary Dugan; Lady Be Good
1942: Jackass Mail; Panama Hattie; The Powers Girl
1943: Swing Shift Maisie
1946: The Kid from Brooklyn
1947: The Secret Life of Walter Mitty; The Road to Rio
1948: Isn't It Romantic?; The Paleface
1950: Let's Dance
1951: My Favorite Spy
1952: Never Wave at a WAC
1954: Casanova's Big Night
1957: Public Pigeon No. 1
1959: Alias Jesse James

WILLIAM SLAVENS McNUTT

1933: Hell and High Water (w/ Grover Jones)

HENRY MacRAE (1888-   )

1933: Rustler's Round-up

(Serials)

1930: The Indians Are Coming; The Lightning Express; Terry of the Times
1932: The Lost Special

LEE MADDEN

1969: Hell's Angels '69
1970: Angels Unchained
1975: The Night God Screamed; The Manhandlers
1978: Night Creature

BEN MADDOW

1960: The Savage Eye (doc.)
1963: An Affair of the Skin

JOHN MAGNUSON

1974: Lenny Bruce Performance Film (doc.)

PAUL MAGWOOD

1972: Chandler

NORMAN MAILER (1923-)

1968: Wild 90; Beyond the Law
1971: Maidstone

ANTHONY MAJOR

1974: Super Spook

HAL MAKELIM

1953: Man of Conflict

KARL MALDEN (1914-)

1957: Time Limit

TERENCE MALICK (1945-)

1974: Badlands
1978: Days of Heaven

LOUIS MALLE (1932-) Fr.

(American films only)
1978: Pretty Baby

LEO MALONEY

1929: Overland Bound

BRUCE K. MAMEL

1977: The First Nudie Musical (w/ Mack Haggard)

ROUBEN MAMOULIAN (1898-)

1929: Applause
1931: City Streets
1932: Dr. Jekyll and Mr. Hyde; Love Me Tonight
1933: Queen Christina; Song of Songs
1934: We Live Again
1935: Becky Sharp
1936: The Gay Desperado
1937: High, Wide and Handsome
1939: Golden Boy
1940: The Mark of Zorro
1941: Blood and Sand
1942: Rings on Her Fingers
1948: Summer Holiday
1957: Silk Stockings

GEORGE MANASSE

1973: Blade

MILES MANDER (1888-1946)

1932: Fascination; The Woman Decides
1935: The Morals of Marcus
1936: The Flying Doctor

JOE MANDUKE

1971: Jump
1975: Cornbread, Earl and Me

JOSEPH L. MANKIEWICZ (1909-)

1946: Dragonwyck; Somewhere in the Night
1947: The Ghost and Mrs. Muir; The Late George Apley
1948: Escape (Brit.); A Letter to Three Wives
1949: House of Strangers
1950: All About Eve; No Way Out
1951: People Will Talk
1952: Five Fingers
1953: Julius Caesar
1954: The Barefoot Contessa
1955: Guys and Dolls
1958: The Quiet American
1959: Suddenly, Last Summer
1963: Cleopatra
1967: The Honey Pot
1969: Couples; The Bawdy Bard
1970: King: A Filmed Record ... Montgomery to Memphis (doc. w/ Sidney Lumet); There Was a Crooked Man ...
1972: Sleuth

ANTHONY MANN (1906-1967)

1942: Dr. Broadway; Moonlight in Havana
1943: Nobody's Darling
1944: Strangers in the Night; My Best Gal
1945: The Great Flamarion; Sing Your Way Home; Two O'Clock Courage
1946: Strange Impersonation; The Bamboo Blonde
1947: T-Men; Desperate; Railroaded
1948: Raw Deal
1949: Border Incident; Reign of Terror (orig. The Black Book); Side Street
1950: Devil's Doorway; The Furies
1951: The Tall Target; Winchester 73
1952: Bend of the River
1953: The Naked Spur; Thunder Bay
1954: The Glenn Miller Story
1955: The Far Country; The Last Frontier; The Man from Laramie; Strategic Air Command
1956: Serenade
1957: Men in War; The Tin Star
1958: God's Little Acre; Man of the West
1960: Cimarron
1961: El Cid
1964: The Fall of the Roman Empire
1965: The Heroes of Telemark
1968: A Dandy in Aspic (Brit.)

DANIEL MANN (1912-)

1952: Come Back, Little Sheba

1954: About Mrs. Leslie
1955: The Rose Tattoo; I'll Cry Tomorrow
1956: The Teahouse of the August Moon
1958: Hot Spell
1959: The Last Angry Man
1960: The Mountain Road; Butterfield 8
1961: Ada
1962: Five Finger Exercise; Who's Got the Action?
1963: Who's Been Sleeping in My Bed?
1966: Our Man Flint; Judith
1968: For Love of Ivy
1969: A Dream of Kings
1971: Willard
1972: The Revengers
1973: Interval; Maurie
1974: Lost in the Stars
1978: Matilda

DELBERT MANN (1920-)

1955: Marty
1957: The Bachelor Party
1958: Desire Under the Elms; Separate Tables
1959: Middle of the Night
1960: The Dark at the Top of the Stairs
1961: Lover Come Back; The Outsider
1962: That Touch of Mink
1963: A Gathering of Eagles
1964: Dear Heart; Quick, Before It Melts
1966: Mister Buddwing
1967: Fitzwilly
1968: The Pink Jungle
1971: Kidnapped (Brit.)
1976: Birch Interval

EDWARD MANN

1956: Scandal Incorporated
1966: Hallucination Generation
1971: Cauldron of Blood; Who Says I Can't Ride a Rainbow!
1972: Hot Pants Holiday

BRUCE MANNING

1943: The Amazing Mrs. Holliday

DUNCAN MANSFIELD

1936: Along Came Love (w/ B. Lytell)
1937: Girl Loves Boy; Sweetheart of the Navy

PAUL MANSLANSKY

1974: Sugar Hill

PAUL MANTZ

1956: Seven Wonders of the World (doc. w/ A. Marton)

GEORGE MANUPELLI

1971: Cry Dr. Chicago

ALEX MARCH

1968: Paper Lion
1969: The Big Bounce
1976: Mastermind

MAX MARCIN

1931: The Lawyer's Secret (w/ L. Gasnier); Silence (w/ L. Gasnier)
1932: The Strange Case of Clara Deane (w/ L. Gasnier)
1933: King of the Jungle (w/ H. B. Humberstone); Gambling Ship (w/ L. Gasnier)
1934: The Love Captive

EDWIN L. MARIN (1901-1951)

1933: The Death Kiss; A Study in Scarlet; The Avenger; The Sweetheart of Sigma Chi
1934: Bombay Mail; The Crosby Case; Affairs of a Gentleman; Paris Interlude
1935: The Casino Murder Case; Pursuit
1936: Moonlight Murder; Speed; Sworn Enemy; I'd Give My Life; All American Chump; The Garden Murder Case
1937: Man of the People; Married Before Breakfast
1938: Everybody Sing; Hold That Kiss; The Chaser; Listen, Darling; A Christmas Carol
1939: Fast and Loose; Society Lawyer; Maisie
1940: Henry Goes Arizona; Florian; Gold Rush Maisie; Hullabaloo
1941: Maisie Was a Lady; Ringside Maisie; Paris Calling
1942: A Gentleman After Dark; Miss Annie Rooney; Invisible Agent
1943: Two Tickets to London
1944: Show Business; Tall in the Saddle
1945: Johnny Angel
1946: Abiline Town; Young Widow; Lady Luck; Mr. Ace; Nocturne
1947: Christmas Eve; Intrigue
1948: Race Street

1949: Canadian Pacific; The Younger Brothers; Fighting Man of the Plains
1950: Colt .45; The Cariboo Trail
1951: Sugarfoot; Raton Pass; Fort Worth

PETER MARIS

1979: Delirium

FLETCHER MARKLE

1949: Jigsaw
1951: The Man with a Cloak; Night into Morning
1963: The Incredible Journey

ROBERT MARKOWITZ

1979: Voices

ARTHUR MARKS

1970: Togetherness
1973: Bonnie's Kids; The Room Mates; Detroit 9000 (w/ S. Latham)
1975: Bucktown; Friday Foster; A Woman for All Men
1976: J. D.'s Revenge; The Monkey Hustle

PAUL MARLOW

1975: Threshold--The Blue Angels Experience (doc.)

RICHARD MARQUAND

1979: The Legacy (Brit.)

RAY MARSH

1974: The Mad, Mad, Movie Makers
1975: Lord Shango

ANTHONY MARSHALL

1943: Bullets and Saddles

DON MARSHALL

1971: Cycles South

GEORGE MARSHALL (1891-   )

1932: Pack Up Your Troubles (w/ Raymond McCarey)
1934: Ever Since Eve; Wild Gold; She Learned About Sailors; 365 Nights in Hollywood
1935: Life Begins at 40; Ten Dollar Raise; In Old Kentucky; Music Is Magic; Show Them No Mercy
1936: A Message to Garcia; The Crime of Dr. Forbes; Can This Be Dixie?
1937: Nancy Steele Is Missing; Love Under Fire
1938: The Goldwyn Follies; The Battle of Broadway; Hold That Co-ed
1939: You Can't Cheat an Honest Man; Destry Rides Again
1940: The Ghost Breakers; When the Daltons Rode
1941: Pot o' Gold; Texas
1942: Valley of the Sun; The Forest Rangers; Star Spangled Rhythm
1943: True to Life; Riding High
1944: And the Angels Sing
1945: Murder, He Says; Incendiary Blonde; Hold That Blonde
1946: The Blue Dahlia; Monsieur Beaucaire
1947: The Perils of Pauline; Variety Girl
1948: Hazard; Tap Roots
1949: My Friend Irma
1950: Fancy Pants; Never a Dull Moment
1951: A Millionaire for Christy
1952: The Savage
1953: Off Limits; Scared Stiff; Houdini; Money from Home
1954: Red Garters; Duel in the Jungle; Destry
1955: The Second Greatest Sex
1956: Pillars of the Sky
1957: The Guns of Fort Petticoat; Beyond Mombasa (Brit.); The Sad Sack
1958: The Sheepman; Imitation General
1959: The Mating Game; It Started with a Kiss; The Gazebo
1961: Cry for Happy
1962: The Happy Thieves; How the West Was Won (w/ others)
1963: Papa's Delicate Condition
1964: Dark Purpose; Advance to the Rear
1966: Boy, Did I Get a Wrong Number!
1967: Eight on the Lam
1968: The Wicked Dreams of Paula Schultz
1969: Hook, Line and Sinker

WILLIAM MARSHALL (1917-   )

1951: Adventures of Capt. Fabian
1961: The Phantom Planet

ALPHONSE MARTELL

1933: Gigolettes of Paris

CHARLES MARTIN

1945: No Leave, No Love
1948: My Dear Secretary
1956: Death of a Scoundrel
1968: If He Hollers, Let Him Go
1974: How to Seduce a Woman
1978: One Man Jury
1979: Dead on Arrival

EUGENE MARTIN

1971: Bad Man's River

FRANCIS MARTIN

1933: Tillie and Gus

PAUL MARTIN

1934: Orient Express

RICHARD MARTIN

1973: The Bengal Tiger

LESLIE H. MARTINSON

1954: The Atomic Kid
1956: Hot Rod Girl
1957: Hot Rod Rumble
1962: Lad: A Dog (w/ Aram Avakian)
1963: PT-109; Black Gold
1964: For Those Who Think Young; FBI Code 98
1966: Batman
1967: Fathom
1971: Mrs. Pollifax--Spy
1976: Escape from Angola

ANDREW MARTON (1904-   )

1935: Miss President (Hung.)
1936: Wolf's Clothing (Brit.)

1939: School for Husbands (Brit.)
1940: A Little Bit of Heaven
1944: Gentle Annie
1946: Gallant Bess
1950: King Solomon's Mines (w/ Compton Bennett; Africa)
1952: Storm over Tibet; The Wild North; The Devil Makes Three (Germ. and Austria)
1954: Gypsy Colt; Prisoner of War; Men of the Fighting Lady; Green Fire (Africa)
1956: Seven Wonders of the World (w/ others)
1958: Underwater Warrior
1962: It Happened in Athens (Gr.); The Longest Day (w/ Ken Annakin, Bernhard Wicki; Fr.)
1964: The Thin Red Line (Span.)
1965: Clarence, the Cross-Eyed Lion; Crack in the World
1966: Around the World Under the Sea; Birds Do It
1967: Africa--Texas Style! (Kenya)

JOSEPH MASCELLI

1964: Monstrosity

NOEL MASON (see NOEL MASON SMITH)

JOE MASSOT

1976: The Song Remains the Same (doc. w/ Peter Clifton)

QUENTIN MASTERS

1972: Thumb Tripping

RUDOLPH MATE (1898-1964) Pol.

1947: It Had to Be You (w/ Don Hartman)
1949: The Dark Past
1950: No Sad Songs for Me; Union Station; Branded
1951: The Prince Who Was a Thief; When Worlds Collide
1952: The Green Glove; Paula; Sally and Saint Anne
1953: The Mississippi Gambler; Second Chance; Forbidden
1954: The Siege at Red River; The Black Shield of Falworth
1955: The Violent Men; The Far Horizons
1956: Miracle in the Rain; The Rawhide Years; Port Afrique (Brit.); Three Violent People
1958: The Deep Six
1959: For the First Time (Ital.)
1962: The 300 Spartans (Gr.)
1963: Aliki--My Love (Gr.); Seven Seas to Calais (Ital.)

ERNEST MATRAY

1944: Adventure in Music (w/ others)

ALEX MATTER

1967: The Drifter

WALTER MATTHAU (1920-)

1960: Gangster Story

FRANK S. MATTISON

1929: Broken Hearted; Bye-Bye Buddy

NORMAN MAURER

1963: The Three Stooges Go Around the World in a Daze
1965: The Outlaws Is Coming

J. P. MAWRA

1965: All Men Are Apes

ELAINE MAY (1932-)

1971: A New Leaf
1972: The Heartbreak Kid
1976: Mikey and Nicky

JOE MAY (1880-1954) Austrian

(American films only)
1934: Music in the Air
1936: One Hour of Romance
1937: Confession
1939: House of Fear; Society Smugglers
1940: The House of the Seven Gables; The Invisible Man Returns; You're Not So Tough
1941: Hit the Road
1944: Johnny Doesn't Live Here Any More

RUSS MAYBERRY

1971: The Jesus Trip
1979: Unidentified Flying Oddball

GERALD MAYER (1919-)

1950: Dial 1119
1951: Inside Straight; The Sellout
1952: Holiday for Sinners
1953: Bright Road
1955: The Marauders
1958: Diamond Safari

KEN MAYNARD (1895-1973)

1933: Fiddlin' Buckaroo

ARCHIE L. MAYO (1891-1968)

1929: My Man; Sonny Boy; The Sap; Is Everybody Happy?; The Sacred Flame
1930: Vengeance; Wide Open; Courage; Oh, Sailor, Behave!; The Doorway to Hell
1931: Illicit; Svengali; Bought
1932: Under Eighteen; The Expert; Street of Women; Two Against the World; Night After Night
1933: The Life of Jimmy Dolan; The Mayor of Hell; Ever in My Heart; Convention City
1934: Gambling Lady; The Man with Two Faces; Desirable
1935: Bordertown; Go into Your Dance; The Case of the Lucky Legs
1936: The Petrified Forest; I Married a Doctor; Give Me Your Heart; Black Legion
1937: Call It a Day; It's Love I'm After
1938: The Adventures of Marco Polo; Youth Takes a Fling
1939: They Shall Have Music
1940: The House Across the Bay; Four Sons
1941: The Great American Broadcast; Charley's Aunt; Confirm or Deny
1942: Moontide; Orchestra Wives
1943: Crash Dive
1944: Sweet and Lowdown
1946: A Night in Casablanca; Angel on My Shoulder

ALBERT AND DAVID MAYSLES

1970: Gimme Shelter (doc.)
1976: Grey Gardens (doc.)

PAUL MAZURSKY

1969: Bob & Carol & Ted & Alice
1970: Alex in Wonderland
1973: Blume in Love
1974: Harry and Tonto
1976: Next Stop, Greenwich Village
1978: An Unmarried Woman
1979: Willie and Paul

DON MEDFORD

1966: To Trap a Spy
1971: The Hunting Party; The Organization

GRAHAM MEECH-BURKESTONE

1979: Burnout

RUDOLPH MEINERT

1930: Strange Case of District Attorney M

GUS MEINS (-1940)

1934: Babes in Toyland (w/ Charles Rogers)
1936: Kelly the Second; Nobody's Baby; The Hit Parade
1937: The Californians; Roll Along, Cowboy
1938: Ladies in Distress; The Higgins Family; His Exciting Night; Romance on the Run
1939: My Wife's Relatives; Money to Burn; The Mysterious Miss X; The Covered Trailer; Should Husbands Work?
1940: Earl of Puddlestone; Grandpa Goes to Town; Scatterbrain

ADOLFAS MEKAS

1964: Guns of the Trees (w/ J. Mekas); The Brig (w/ J. Mekas)
1968: Windflowers

JONAS MEKAS (1922-)

1964: Guns of the Trees; The Brig (both w/ A. Mekas)
1972: Reminiscences of a Journey to Lithuania (doc.)

IB MELCHIOR (1917-)

1959: The Angry Red Planet
1964: The Time Travelers

BILL MELENDEZ

1969: A Boy Named Charlie Brown (anim.)
1972: Snoopy, Come Home (anim.)
1977: Race for Your Life, Charlie Brown (anim.)

GEORGE MELFORD (1877-1961)

1929: Love in the Desert; Sea Fury
1930: The Poor Millionaire
1931: East of Borneo; Homicide Squad (w/ E. L. Cahn); The Viking
1932: The Boiling Point; The Cowboy Counsellor; A Scarlet Week-End; The Penal Code
1933: The Dude Bandit; Man of Action; Officer 13; Eleventh Commandment
1934: Hired Wife
1935: East of Java

(Serial)

1937: Jungle Menace (w/ Harry Fraser)

LOTHAR MENDES (1894-1974) Germ.

1929: Interference (w/ Roy Pomeroy); The Four Feathers (w/ Merian C. Cooper, E. B. Schoedsack); Illusion; Marriage Playground
1930: Paramount on Parade (w/ others)
1931: Ladies' Man
1932: Strangers in Love; Payment Deferred
1933: Luxury Liner
1934: Power (Brit.; Jew Suss)
1937: The Man Who Could Work Miracles (Brit.)
1938: Moonlight Sonata (Brit.)
1941: International Squadron
1943: Flight for Freedom
1944: Tampico
1946: The Walls Came Tumbling Down

GIAN-CARLO MENOTTI (1911-) Ital./U.S.

1951: The Medium

WILLIAM CAMERON MENZIES (1896-1957)

1931: Always Goodbye (w/ Kenneth McKenna); The Spider (w/ K. McKenna)
1932: Almost Married; Chandu the Magician (w/ Marcel Varnel)
1933: I Love You Wednesday (w/ H. King)
1934: Wharf Angel (w/ George Somnes)
1936: Things to Come (Brit.)
1937: Four Dark Hours (ret. The Green Cockatoo; Brit.)
1944: Address Unknown
1951: Drums in the Deep South; The Whip Hand
1953: The Maze; Invaders from Mars

BURGESS MEREDITH (1908-   )

1949: The Man on the Eiffel Tower

RON MERK

1979: Pinocchio's Storybook Adventures

LAURENCE MERNICH

1973: Manson (doc.)

GEORGE M. MERRICK

1933: Secrets of Hollywood
1937: Angkor; or Forbidden Adventure in Angkor
1942: Today I Hang (w/ Oliver Drake)

KEITH MERRILL

1974: The Great American Cowboy (doc.)
1978: Take Down

PHILIP F. MESSINA

1970: Skezag (doc. w/ Joel F. Freedman)

RADLEY H. METZGER (1930-   )

1961: Dark Odyssey (w/ William Kyriakys)
1964: The Dirty Girls
1966: The Alley Cats
1967: Carmen Baby (Yugoslavia/Germ./U.S.)

1968: Therese and Isabelle (Fr.)
1969: Camille 2000 (Ital.)
1970: The Lickerish Quartet (U.S./Germ./Ital.)
1972: Little Mother (U.S./Yugoslavia)
1973: Score (U.S./Yugoslavia)

ANDREW MEYER

1970: The Sky Pirate
1974: Night of the Cobra Woman

HERBERT MEYER

1939: Bad Boy; Son of Ingagi (w/ Richard Kahn)

NICHOLAS MEYER

1979: Time After Time

RUSS MEYER (c.1924-   )

1959: The Immoral Mr. Teas
1961: Eroticon; Eve and the Handyman
1962: Naked Gals of the Golden West
1963: Europe in the Raw (Europe); Heavenly Bodies
1964: Lorna
1965: Mudhoney!; Motor Psycho!; Fanny Hill (Germ.)
1966: Faster Pussycat!; Kill! Kill!; Mondo Topless
1967: Good Morning--and Goodbye; Common Law Cabin
1968: Finders Keepers, Lovers Weepers; Russ Meyer's Vixen
1969: Cherry, Harry and Raquel
1970: Beyond the Valley of the Dolls
1971: The Seven Minutes
1973: Blacksnake! (Sweet Suzy)
1975: The Supervixens

SIDNEY MEYERS (1906-1969)

1949: The Quiet One (doc.)
1959: The Savage Eye (doc. w/ others)

RICHARD MICHAELS

1975: How Come Nobody's on Our Side?
1976: Death Is Not the End (doc.)

OSCAR MICHEAUX (-1951)

1929: Wages of Sin
1930: Daughter of the Congo; Easy Street
1931: Darktown Review; The Exile
1932: Black Magic; Ten Minutes to Live; Veiled Aristocrats
1933: The Girl from Chicago; Ten Minutes to Kill
1934: Harlem After Midnight
1935: Lem Hawkin's Confession
1936: Temptation; Underworld
1937: God's Stepchildren
1948: The Betrayal

GEORGE MIDDLETON

1930: Double Cross Roads (w/ A. L. Werker)

TED V. MIKELS

1972: The Corpse Grinders
1974: The Doll Squad

WILLIAM MILES

1977: Men of Bronze (doc.)

LEWIS MILESTONE (1895-1980)

1929: Betrayal
1930: All Quiet on the Western Front
1931: The Front Page
1932: Rain
1933: Hallelujah, I'm a Bum
1934: The Captain Hates the Sea
1935: Paris in Spring
1936: Anything Goes; The General Died at Dawn
1939: Of Mice and Men; The Night of Nights
1940: Lucky Partners
1941: My Life with Caroline
1943: Edge of Darkness; The North Star
1944: The Purple Heart
1945: A Walk in the Sun
1946: The Strange Love of Martha Ivers
1948: Arch of Triumph; No Minor Vices
1949: The Red Pony
1950: Halls of Montezuma
1952: Kangaroo; Les Miserables
1953: Melba (Brit.); They Who Dare (Brit.)
1955: The Widow (Ital.)

1959: Pork Chop Hill
1960: Ocean's Eleven
1961: Mutiny on the Bounty

GENE MILFORD

1960: The Pusher

JOHN MILIUS

1973: Dillinger
1975: The Wind and the Lion
1978: Big Wednesday

WARREN MILLAIS

1933: Her Secret

RAY MILLAND (1905-)

1955: A Man Alone
1956: Lisbon (Port.)
1958: The Safecracker
1962: Panic in the Year Zero
1968: Hostile Witness (Brit.)

STUART MILLAR (1929-)

1972: When the Legends Die
1975: Rooster Cogburn

DAVID MILLER (1909-)

1941: Billy the Kid
1942: Flying Tigers; Sunday Punch
1949: Love Happy; Top o' the Morning
1950: Our Very Own
1951: Saturday's Hero
1952: Sudden Fear
1954: Twist of Fate (Brit.)
1955: Diane
1956: The Opposite Sex
1957: The Story of Esther Costello (Brit.)
1959: Happy Anniversary
1960: Midnight Lace
1961: Back Street
1962: Lonely Are the Brave

1963: Captain Newman, M.D.
1968: Hammerhead
1969: Hail, Hero!
1973: Executive Action
1977: Bittersweet Love

GILBERT MILLER

1933: The Lady Is Willing

IRA MILLER

1979: Coming Attractions

MICHAEL MILLER

1975: Sweet Girls
1976: Jackson County Jail

ROBERT ELLIS MILLER (1927-)

1966: Any Wednesday
1968: The Heart Is a Lonely Hunter; Sweet November
1971: The Buttercup Chain
1974: The Girl from Petrovka

SIDNEY MILLER

1959: The 30-Foot Bride of Candy Rock
1964: Get Yourself a College Girl
1967: Tammy and the Millionaire (w/ Leslie Goodwins)

WARREN MILLER

1967: Ski on the Wild Side (doc.)

DAN MILNER

1957: From Hell It Came

ROBERT MILTON (c.1890-)

1929: The Dummy

1930: Behind the Makeup; Outward Bound
1931: The Bargain; Devotion
1932: Westward Passage; Husband's Holiday
1933: Dance of Witches (Brit.)
1934: Bella Donna (Brit.)

MICHAEL MINDLIN

1934: Hitler's Reign of Terror (doc.)

ALLEN H. MINER

1955: The Naked Sea (doc.)
1956: Ghost Town
1957: Black Patch; The Ride Back
1968: Chubasco

WORTHINGTON MINER

1934: Let's Try Again; Hat, Coat and Glove

VINCENTE MINNELLI (1910-)

1943: I Dood It; Cabin in the Sky
1944: Meet Me in St. Louis
1945: The Clock; Yolanda and the Thief
1946: Undercurrent; Ziegfeld Follies
1948: The Pirate
1949: Madame Bovary
1950: Father of the Bride
1951: An American in Paris; Father's Little Dividend
1952: The Bad and the Beautiful
1953: The Band Wagon; Home at Seven (Brit.); The Story of Three Lovers (w/ G. Reinhardt)
1954: Brigadoon; The Long, Long Trailer
1955: The Cobweb; Kismet
1956: Lust for Life; Tea and Sympathy
1957: Designing Woman
1958: Gigi; The Reluctant Debutante; Some Came Running
1960: Bells Are Ringing; Home from the Hill
1962: Four Horsemen of the Apocalypse; Two Weeks in Another Town
1963: The Courtship of Eddie's Father
1964: Goodbye Charlie
1965: The Sandpiper
1970: On a Clear Day You Can See Forever
1976: A Matter of Time

BRUCE MITCHELL

1930: The Lonesome Trail
1931: Sheer Luck
1932: 45 Calibre Echo
1934: The Rainbow Terror

LUKE MOBERLY

1973: Little Laura and Big John (w/ Bob Woodburn)

PHILIP MOELLER

1934: The Age of Innocence
1935: Break of Hearts

LEONIDE MOGUY (1899-1976)

(American films only)
1943: Paris After Dark
1944: Action in Arabia
1946: Whistle Stop

HAL MOHR (1894-1974)

1937: When Love Is Young

DAVID MONAHAN

1970: The Phantom Tollbooth (w/ others)

BENJAMIN MONASTER

1965: Goldstein (w/ Philip Kaufman)

JOHN MONKS, JR.

1962: No Man Is an Island (w/ Richard Goldstone)

EDWARD J. MONTAGNE

1949: Project X
1950: The Tattooed Stranger

1951: The Man with My Face
1964: McHale's Navy
1965: McHale's Navy Joins the Air Force
1967: The Reluctant Astronaut
1978: They Went That-A-Way and That-A-Way

GEORGE MONTGOMERY (1916-)

1961: The Steel Claw
1962: Samar
1970: Satan's Harvest

ROBERT MONTGOMERY (1904-)

1946: Lady in the Lake
1947: Ride the Pink Horse
1949: Once More My Darling
1950: Eye Witness (Brit.)
1960: The Gallant Hours

TITUS MOODY

1974: The Last of the American Hoboes

JAMES MOORE

1940: The Secret Seven

MICHAEL MOORE

1966: An Eye for an Eye; Paradise, Hawaiian Style
1967: Kill a Dragon
1968: Buckskin; The Fastest Guitar Alive

RICHARD MOORE

1979: Circle of Iron

ROBERT MOORE

1976: Murder by Death
1978: The Cheap Detective
1979: Chapter Two

VIN MOORE

1930: Cohens and Kellys in Africa
1931: Virtuous Husband; Many a Slip; Ex-Bad Boy
1932: Racing Youth
1934: Love Past Thirty; Flirting with Danger
1935: Cheers of the Crowd
1936: The Dragnet
1938: Topa Topa (w/ Charles Hutchison)
1940: Killers of the Wild

WILLIAM MORGAN

1941: Mr. District Attorney; Sierra Sue; Sunset in Wyoming; Mercy Island; Bowery Boy; The Gay Vagabond
1942: Bells of Capistrano; Cowboy Serenade; The Heart of the Rio Grande; Home in Wyomin'; Stardust on the Sage; Secrets of the Underground
1943: Headin' for God's Country
1947: Fun and Fancy Free

DAVID BURTON MORRIS

1975: Loose Ends

HOWARD MORRIS (1919-)

1967: Who's Minding the Mint?
1968: With Six You Get Egg Roll
1969: Don't Drink the Water
1978: Goin' Coconuts

PAUL MORRISSEY (1939-)

1968: Flesh
1970: Trash
1974: Andy Warhol's Frankenstein; Andy Warhol's Dracula

VIC MORROW (1932-)

1967: Deathwatch
1971: A Man Called Sledge (Ital.)

HOLLINGSWORTH MORSE

1970: Puffnstuff
1972: Daughters of Satan

TERRY O. MORSE (1906-   )

1939: Adventures of Jane Arden; On Trial; Waterfront; Smashing the Money Ring; No Place to Go
1940: British Intelligence; Tear Gas Squad
1945: Fog Island
1946: Danny Boy; Shadows over Chinatown; Dangerous Money
1947: Bells of San Fernando
1951: Unknown World
1956: Godzilla, King of the Monsters (w/ Ishiro Honda; Jap.)
1965: Taffy and the Jungle Hunter; Young Dillinger

GILBERT MOSES

1973: Willie Dynamite
1979: The Fish That Saved Pittsburgh

JACK MOSS

1945: Snafu

SARKY MOURADIAN

1974: Tears of Happiness

DIETER MULLER

1970: The Last Mercenary

ROBERT MULLIGAN (1925-   )

1957: Fear Strikes Out
1960: The Great Imposter; The Rat Race
1961: Come September
1962: The Spiral Road; To Kill a Mockingbird
1963: Love with the Proper Stranger
1965: Baby, the Rain Must Fall; Inside Daisy Clover
1967: Up the Down Staircase
1969: The Stalking Moon; The Piano Sport
1971: The Pursuit of Happiness; Summer of '42
1972: The Other

1974: The Nickel Ride
1978: Bloodbrothers; Same Time, Next Year

CHRIS MUNGER

1972: Kiss of the Tarantula
1974: Black Starlet

F. W. MURNAU (1889-1931) Germ.

(American films only)
1929: Four Devils
1930: Our Daily Bread (ret. City Girl)
1931: Tabu (w/ Robert Flaherty)

DUDLEY MURPHY (1897-   )

1931: Confessions of a Co-Ed (w/ David Burton)
1932: The Sport Parade
1933: The Emperor Jones
1935: The Night Is Young
1936: Don't Gamble with Love
1939: One Third of a Nation; Main Street Lawyer

PATRICK MURPHY

1972: Squares

RALPH MURPHY (1895-1967)

1931: The Big Shot
1932: 70,000 Witnesses; Panama Flo
1933: Strictly Personal; Song of the Eagle; Girl Without a Room; Golden Harvest
1934: The Notorious Sophie Lang; She Made Her Bed; Private Scandal; The Great Flirtation; Menace
1935: McFadden's Flats; Men Without Names; One Hour Late
1936: Florida Special; Collegiate; The Man I Marry; Top of the Town
1937: Night Club Scandal; Partners in Crime
1939: Our Neighbors--The Carters
1940: I Want a Divorce
1941: You're the One; Las Vegas Nights; Glamour Boy; Midnight Angel
1942: Mrs. Wiggs of the Cabbage Patch; Pacific Blackout
1943: Night Plane from Chungking
1944: The Man in Half Moon Street; Rainbow Island; The Town Went Wild

1945: Sunbonnet Sue; How Do You Do?
1947: The Spirit of West Point
1948: Mickey
1949: Red Stallion in the Rockies
1951: The Lady and the Bandit; Never Trust a Gambler; Stage to Tucson
1952: Lady in the Iron Mask; Captain Pirate
1955: Mystery of the Black Jungle (Germ.)

RICHARD MURPHY (1912-)

1955: Three Stripes in the Sun
1960: The Wackiest Ship in the Army

DON MURRAY (1929-)

1972: The Cross and the Switchblade

FLOYD MUTRUX

1971: Dusty and Sweets McGee (doc.)
1975: Aloha, Bobby and Rose
1978: American Hot Wax

ZION MYERS

1931: Sidewalks of New York (w/ J. White)
1933: Lucky Dog

ALAN MYERSON

1973: Steelyard Blues

ARTHUR NADEL

1967: Clambake

CONRAD NAGEL (1896-1970)

1937: Love Takes Flight

IVAN NAGY

1973: Bad Charleston Charlie
1976: Deadly Hero

MICHAEL NAHAY

1976: The Thursday Morning Murders

ARTHUR NAPOLEON

1957: Man on the Prowl
1958: Too Much, Too Soon
1969: The Activist

GENE NASH

1970: Dinah East

EDWARD NASSOUR

1956: The Beast of Hollow Mountain (w/ I. Rodriguez)

WILLIAM T. NAUD

1966: Hot Rod Hullabaloo
1972: Wild in the Sky (ret. Black Jack)

GREGORY NAVA

1977: The Confessions of Amans

CAL NAYLOR

1979: Dirt (w/ Eric Karson)

RAY NAZARRO (1902-)

1945: Outlaws of the Rockies; Texas Panhandle
1946: Desert Horseman; Galloping Thunder; Gunning for Vengeance; Headin' West; Roaring Rangers; Terror Trail; Two Fisted Stranger; That Texas Jamboree; Throw a Saddle on a Star; Cowboy Blues; Singing on the Trail; Lone Star Moonlight
1947: Buckaroo from Powder River; Last Days of Boot Hill; Law of the Canyon; The Lone Hand Texan; West of Dodge City; Over the Santa Fe Trail; Rose of Santa Rosa
1948: El Dorado Pass; Phantom Valley; Singing Spurs; Six-Gun Law; Trail to Laredo; West of Sonora; Song of Idaho; Smoky Mountain Melody
1949: Bandits of El Dorado; The Blazing Trail; Challenge of the Range; Laramie; Quick on the Trigger; Renegades of the Sage; South of Death Valley

1950: The Tougher They Come; Frontier Outpost; Outcast of Black Mesa; The Palomino; Streets of Ghost Town; Texas Dynamo; Trail of the Rustlers; David Harding--Counterspy; Hoedown
1951: Al Jennings of Oklahoma; China Corsair; Flame of Stamboul; Cyclone Fury; Fort Savage Raiders; The Kid from Amarillo; War Cry
1952: Cripple Creek; Indian Uprising; Montana Territory; Junction City; Laramie Mountains; The Rough Tough West
1953: The Bandits of Corsica; Gun Belt; Kansas Pacific
1954: The Black Dakotas; The Lone Gun; Southwest Passage
1955: Top Gun
1956: The White Squaw
1957: The Domino Kid; The Hired Gun; The Phantom Stagecoach
1958: Apache Territory; Return to Warbow

RONALD NEAME (1911-) Brit.

(American films only)
1962: Escape from Zahrain
1965: Mister Moses
1966: Gambit; A Man Could Get Killed
1972: The Poseidon Adventure
1974: The Odessa File
1979: Meteor

HAL NEEDHAM

1977: Smokey and the Bandit
1978: Hooper
1979: The Villain

JEAN NEGULESCO (1900-)

1941: Singapore Woman
1944: The Conspirators; The Mask of Dimitrios
1946: Humoresque; Nobody Lives Forever; Three Strangers
1947: Deep Valley
1948: Johnny Belinda; Road House
1949: The Forbidden Street
1950: The Mudlark; Three Came Home; Under My Skin
1951: Take Care of My Little Girl
1952: Lure of the Wilderness; Lydia Bailey; Phone Call from a Stranger; O. Henry's Full House (w/ others)
1953: How to Marry a Millionaire; Scandal at Scourie; Titanic
1954: Three Coins in the Fountain; Woman's World
1955: Daddy Long Legs; The Rains of Ranchipur
1957: Boy on a Dolphin
1958: A Certain Smile; The Gift of Love
1959: The Best of Everything; Count Your Blessings
1962: Jessica

1964: The Pleasure Seekers
1970: Hello-Goodbye (Brit.); The Invincible Six

MARSHALL NEILAN (1891-1958)

1929: The Vagabond Lover; Tanned Legs; The Awful Truth; Taxi 13
1930: Sweethearts on Parade
1934: The Lemon Drop Kid; Social Register; Chloe
1935: This Is the Life
1937: Sing While You're Able; Swing It, Professor; Thanks for Listening

ROY WILLIAM NEILL (1890-1946)

1929: Behind Closed Doors; Wall Street
1930: The Melody Man; Cock o' the Walk (w/ Walter Lang); Just Like Heaven
1931: The Avenger; The Good Bad Girl; Fifty Fathoms Deep
1932: The Menace; That's My Boy
1933: The Circus Queen Murder; The Whirlpool; As the Devil Commands; Above the Clouds
1934: Fury of the Jungle; The Ninth Guest; Black Moon; Blind Date; I'll Fix It; Jealousy
1935: Mills of the Gods; Eight Bells; The Black Room
1936: The Lone Wolf Returns
1937: Dr. Syn (Brit.)
1938: Thank Evans (Brit.); Simply Terrific (Brit.); The Viper (Brit.); Everything Happens to Me (Brit.); Many Tanks Mr. Atkins (Brit.)
1939: The Good Old Days (Brit.); Murder Will Out (Brit.); Hoots Mon! (Brit.); His Brother's Keeper (Brit.)
1942: Madame Spy; Sherlock Holmes and the Secret Weapon
1943: Eyes of the Underworld; Frankenstein Meets the Wolf Man; Rhythm of the Islands; Sherlock Holmes in Washington; Sherlock Holmes Faces Death
1944: Sherlock Holmes and the Spider Woman; The Scarlet Claw; Gypsy Wildcat; The Pearl of Death
1945: The House of Fear; The Woman in Green; Pursuit to Algiers
1946: Black Angel; Dressed to Kill; Terror by Night

JAMES NEILSON (1918-)

1957: Night Passage
1962: Moon Pilot; Bon Voyage!
1963: Summer Magic; Dr. Syn (Brit.)
1964: The Moonspinners (Crete/Brit.)
1966: Return of the Gunfighter
1967: The Adventures of Bullwhip Griffin; Gentle Giant

1968: Where Angels Go--Trouble Follows!
1969: The First Time; Flareup

ALVIN J. NEITZ

1930: Firebrand Jordan; Breed of the West; Trails of Peril
1931: Pueblo Terror; Hell's Valley; Lariats and Six Shooters; Red Fork Range; Flying Lariats
1932: Tex Takes a Holiday

DAVID NELSON

1972: Confessions of Tom Harris (w/ John Derek)

GARY NELSON

1972: Molly and Lawless John
1973: Santee
1977: Freaky Friday
1979: The Black Hole

GENE NELSON (1920-)

1962: Hand of Death
1963: Hootenanny Hoot
1964: Kissin' Cousins; Your Cheatin' Heart
1965: Harum Scarum
1967: The Cool Ones

JACK NELSON

1931: Two Gun Caballero
1934: Border Guns; The Border Menace

MERVYN NELSON

1971: Some of My Best Friends Are...

OZZIE NELSON (1907-1975)

1965: Love and Kisses

RALPH NELSON (1916-)

1962: Requiem for a Heavyweight

1963: Lilies of the Field; Soldier in the Rain
1964: Fate Is the Hunter; Father Goose
1965: Once a Thief
1966: Duel at Diablo
1968: Counterpoint; Charly
1970: Tick ... Tick ... Tick; Soldier Blue
1971: Flight of the Doves (Brit.)
1972: The Wrath of God
1975: The Wilby Conspiracy (Kenya)
1976: Embryo
1978: A Hero Ain't Nothin' but a Sandwich

SAM NELSON

1937: Outlaws of the Prairie
1938: Cattle Raiders; The Colorado Trail; Law of the Plains; Rio Grande; South of Arizona; West of Cheyenne; West of Santa Fe
1939: Parents on Trial; Man from Sundown; North of the Yukon; The Stranger from Texas; Texas Stampede; The Thundering West; Western Caravans
1940: Pioneers of the Frontier; Prairie Schooners; Bullets for Rustlers
1941: Outlaws of the Panhandle
1942: Sagebrush Law
1943: The Avenging Rider

(Serials)

1938: Great Adventures of Wild Bill Hickok (w/ M. V. Wright)
1939: Mandrake, the Magician (w/ Norman Deming); Overland with Kit Carson (w/ N. Deming)

LEWIS NEUMAN

1931: Ubangi

KURT NEUMANN (1908-1958) Germ.

1932: Fast Companions (alt. Information Kid); My Pal, the King
1933: The Big Cage; The Secret of the Blue Room; King for a Night
1934: Let's Talk It Over; Half a Sinner; Wake Up and Dream
1935: Alias Mary Dow; The Affair of Susan
1936: Let's Sing Again; Rainbow on the River
1937: Espionage; Make a Wish; Hold 'Em Navy
1938: Wide Open Faces; Touchdown, Army
1939: Unmarried; Island of Lost Men; Ambush; All Women are Saints
1940: A Night at Earl Carroll's; Ellery Queen--Master Detective

1942: Brooklyn Orchid; About Face; The McGuerins from Brooklyn; Fall In; Taxi, Mister
1943: Yanks Ahoy; The Unknown Guest
1945: Tarzan and the Amazons
1946: Tarzan and Leopard Woman
1947: Tarzan and the Huntress
1948: The Dude Goes West; Bad Men of Tombstone
1949: Bad Boy
1950: The Kid from Texas; Rocketship X-M
1951: Cattle Drive; Reunion in Reno
1952: Son of Ali Baba; The Ring; Hiawatha
1953: Tarzan and the She-Devil
1955: They Were So Young
1956: Mohawk; The Desperadoes Are in Town
1957: She Devil; Kronos; The Deerslayer
1958: The Fly; Circus of Love (Germ.); Machete
1959: Watusi; Counterplot (Puerto Rico)

PETER NEVARD

1970: Groupies (doc. w/ R. Dorfman)

SAM NEWFIELD (SAMUEL NEUFELD) (1900-1964)

1933: Reform Girl; Important Witness; Under Secret Orders
1934: Big Time or Bust; Marrying Widows; Beggar's Holiday
1935: Code of the Mounted; Northern Frontier; Trails of the Wild; Racing Luck; Bulldog Courage
1936: Timber War; Federal Agent; Burning Gold; Border Caballero; Lightnin' Bill Carson; Roarin' Guns; The Lion's Den; Ghost Patrol; Aces and Eights; Go-Get-'Em Haines; The Traitor; Stormy Trails
1937: Melody of the Plains; Doomed at Sundown; Bar Z Bad Man; Roarin' Lead (w/ M. V. Wright); Guns in the Dark; Gun Lords of Stirrup Basin; A Lawman Is Born; Boothill Brigade; Arizona Gunfighter; Ridin' the Lone Trail; The Colorado Kid
1938: Paroled--To Die; Rangers Roundup; Harlem on the Prairie; Code of the Rangers; Six Gun Trail; Thunder in the Desert; Songs and Bullets; Desert Patrol; The Phantom Ranger; Terror of Tiny Town; Frontier Scout; Lightning Carson Rides Again; Crashin' Thru Danger; Lightnin' Crandall; Durango Valley Raiders; The Feud Maker
1939: Six Gun Rhythm; Trigger Fingers; Trigger Pals; Code of the Cactus; Texas Wildcats; Outlaws Paradise; The Fighting Renegade; Fighting Mad; Flaming Lead; The Sagebrush Family Trails West; The Invisible Killer
1940: Beast of Berlin (alt. Goose Step; Hell's Devils); Straight Shooter; Secrets of a Model; Hold That Woman!; I Take This Oath; A Fugitive from Justice; Marked Men; Arizona Gang

Busters; Billy the Kid in Texas; Billy the Kid Outlawed; Billy the Kid's Gun Justice; Frontier Crusaders; Gun Code; Riders of Black Mountain; Texas Renegades

1941: The Lone Rider Ambushed; The Lone Rider Crosses the Rio; The Lone Rider Fights Back; The Lone Rider in Frontier Fury; The Lone Rider in Ghost Town; The Lone Rider Rides On; Billy the Kid Is Wanted; Billy the Kid's Fighting Pals; Billy the Kid's Roundup; Billy the Kid in Santa Fe; Billy the Kid's Range War; Outlaws of the Rio Grande; The Texas Marshal

1942: The Lone Rider and the Bandit; The Lone Rider in Cheyenne; Along the Sundown Trail; Billy the Kid Trapped; Billy the Kid's Smoking Guns; Law and Order; Raiders of the West; Rolling Down the Great Divide; Jungle Siren; The Lone Rider in Border Roundup; The Lone Rider in Texas Justice; The Mad Monster; Outlaws of Boulder Pass; Overland Stagecoach; Prairie Pals; The Mysterious Rider; Sheriff of Sage Valley; Texas Manhunt; Tumbleweed Trail

1943: Queen of Broadway; Western Cyclone; Wild Horse Rustlers; The Kid Rides Again; Danger! Women at Work; Harvest Melody; The Renegade; Tiger Fangs; The Black Raven; Blazing Frontier; Dead Men Walk; Death Rides the Plains; Fugitive of the Plains; Cattle Stampede; Wolves of the Range; Raiders of Red Gap

1944: The Contender; The Drifter; Frontier Outlaws; I Accuse My Parents; Fuzzy Settles Down; The Monster Maker; Nabonga; Oath of Vengeance; Rustler's Hideout; Swing Hostess; Thundering Gun Slingers; Valley of Vengeance; Wild Horse Phantom

1945: Apology for Murder; Border Badmen; Fighting Bill Carson; The Lady Confesses; Gangster's Den; His Brother's Ghost; The Kid Sister; Prairie Rustlers; Stagecoach Outlaws; White Pongo

1946: Lightning Raiders; Outlaw of the Plains; Blonde for a Day; Gashouse Kids; Ghost of Hidden Valley; Lady Chasers; Larceny in Her Heart; Murder Is My Business; Overland Riders; Queen of Burlesque; Terrors on Horseback; The Flying Serpent; Prairie Badmen

1947: Raiders of Red Rock; Three on a Ticket; Frontier Fighters; Code of the Plains;

1948: The Counterfeiters; Money Madness; Lady at Midnight; The Strange Mrs. Crane

1949: State Department File 649; The Devil's Weed (alt. Wild Weed)

1950: Motor Patrol; Radar Secret Service; Western Pacific Agent; Hi-Jacked

1951: The Lost Continent; Skip Along Rosenbloom; Sky High; Fingerprints Don't Lie; Mask of the Dragon; Leave It to the Marines; Three Desperate Men

1952: Outlaw Women (w/ R. Ormond); The Gambler and the Lady; Lady in the Fog

1954: Thunder over Sangoland

1955: Last of the Desperados

1956: Frontier Gambler; The Three Outlaws; The Wild Dakotas (w/ Sig Neufield); Along the Mohawk Trail; The Long Rifle and the Tomahawk (w/ S. Salkow); The Pathfinder and the Mohican; The Redmen and the Renegades
1958: Flaming Frontier; Wolf Dog

JOHN NEWLAND

1957: That Night; The Violators
1966: The Spy with My Face
1974: Legend of Hillbilly John

ANTHONY NEWLEY (1931-) Brit.

(American films only)
1971: Summertree

JOSEPH M. NEWMAN (1909-)

1942: Northwest Rangers
1948: Jungle Patrol
1949: The Great Dan Patch; Abandoned
1950: 711 Ocean Drive
1951: Lucky Nick Cain; The Guy Who Came Back; Love Nest
1952: Red Skies of Montana (alt. Smoke Jumpers); The Outcasts of Poker Flat; Pony Soldier
1953: Dangerous Crossing
1954: The Human Jungle
1955: This Island Earth; Kiss of Fire
1956: Flight to Hong Kong
1957: Death in Small Doses
1958: Fort Massacre
1959: The Gunfight at Dodge City; The Big Circus; Tarzan, the Ape Man
1961: King of the Roaring Twenties--The Story of Arnold Rothstein; A Thunder of Drums; Twenty Plus Two; The George Raft Story (alt. Spin of a Coin)

PAUL NEWMAN (1925-)

1968: Rachel, Rachel
1971: Sometimes a Great Notion (ret. Never Give an Inch)
1972: The Effect of Gamma Rays on Man-in-the-Moon Marigolds

FRED NEWMEYER (1888-)

1929: The Rainbow Man; Sailor's Holiday; It Can Be Done
1930: Fast and Loose; The Grand Parade; Queen High

1931: Subway Express
1932: They Never Come Back; Discarded Lovers; The Fighting Gentleman; Gambling Sex
1933: Easy Millions
1934: The Big Race; The Moth
1935: No Ransom; Secrets of Chinatown
1936: General Spanky (w/ Gordon Douglas)
1942: Rodeo Rhythm

JOEL NEWTON

1953: Jennifer

FRED NIBLO (1874-1948)

1930: Redemption; Way Out West
1931: Donovan's Kid; The Big Gamble
1932: Blame the Woman (Brit.); Two White Arms (Brit.); Diamond Cut Diamond (Brit. w/ M. Elvey)

GEORGE NICHOLLS, JR. (-1940)

1934: Anne of Green Gables; Finishing School (w/ W. Tuchock). The Return of Peter Grimm
1935: Chasing Yesterday
1936: Chatterbox; The Witness Chair; M'Liss; The Big Game
1937: The Soldier and the Lady; Portia on Trial
1938: Army Girl
1939: Man of Conquest
1940: The Marines Fly High (w/ B. Stoloff); High School

CHARLES A. NICHOLS

1973: Charlotte's Web (anim.; w/ others)

DUDLEY NICHOLS (1895-1960)

1943: Government Girl
1946: Sister Kenny
1947: Mourning Becomes Electra

MIKE NICHOLS (1931-)

1966: Who's Afraid of Virginia Woolf?
1967: The Graduate
1970: Catch 22
1971: Carnal Knowledge

1973: The Day of the Dolphin
1975: The Fortune

JACK NICHOLSON (1937-)

1972: Drive, He Said
1978: Goin' South

ALEX NICOL (1919-)

1958: The Screaming Skull
1961: Then There Were Three (Ital.)
1973: Point of Terror

WILLIAM NIGH (1881-)

1929: Desert Nights
1930: Lord Byron of Broadway (w/ H. Beaumont); Fighting Thru; Today
1931: The Single Sin; Lightning Flyer; Sea Ghost
1932: Border Devils; Night Rider; Without Honors
1933: Men Are Such Fools; He Couldn't Take It
1934: City Limits; Monte Carlo Nights; Mystery Liner; Once to Every Bachelor; School for Girls; Two Heads on a Pillow; Without Children; House of Mystery
1935: His Night Out; The Old Homestead; Dizzy Dames; Headline Woman; She Gets Her Man; Mysterious Mr. Wong; Sweepstake Annie
1936: Don't Get Personal; Crash Donovan; North of Nome; Penthouse Party; Steel
1937: The 13th Man; Atlantic Flight; Boy of the Streets; The Hoosier Schoolboy; A Bride for Henry; Bill Cracks Down; The Right to Kill; The Law Commands
1938: Female Fugitive; Rose of the Rio Grande; Gangster's Boy; Mr. Wong, Detective; I Am a Criminal; Romance of the Limberlost
1939: Streets of New York; Mutiny in the Big House; Mr. Wong in Chinatown; The Mystery of Mr. Wong
1940: The Ape; Doomed to Die; The Fatal Hour; Son of the Navy; The Under-dog
1941: The Kid from Kansas; Mob Town; No Greater Sin; Secret Evidence; Zis Boom Bah
1942: Lady from Chungking; Mr. Wise Guy; The Strange Case of Dr. Rx; Black Dragons; City of Silent Men; Tough As They Come; Escape from Hong Kong
1943: Corregidor; Where Are Your Children?; The Ghost and the Guest
1944: Are These Our Parents?; Forever Yours; Trocadero
1945: Allotment Wives; Divorce; They Shall Have Faith
1946: Beauty and the Bandit; The Gay Cavalier; South of Monterey; Partners in Time

1947: Riding the California Trail
1948: I Wouldn't Be in Your Shoes; Stage Struck

JACK NOBEL

1934: At the Sign

WILLIAM NOLTE

1938: Life Goes On

B. L. NORTON

1972: Cisco Pike
1979: More American Graffiti

LLOYD NOSLER

1931: Man from Death Valley
1932: Galloping Thru
1933: Son of the Border

MAX NOSSECK (ALEXANDER M. NORRIS) (1902-1972)

(American films only)
1940: Overture to Glory (Yiddish doc.); Girls Under 21
1941: Gambling Daughters
1945: Dillinger; The Brighton Strangler
1946: Black Beauty
1947: The Return of Rin-Tin-Tin
1950: Kill or Be Killed
1951: Korea Patrol; The Hoodlum
1953: Body Beautiful
1957: Garden of Eden

NOEL NOSSECK

1975: Best Friends
1976: Las Vegas Lady
1978: Youngblood
1979: Dreamer

F. H. NOVIKOW

1974: Blood Couple

SIMON NUCHTERN

1970: Cowards
1972: The Broad Coalition; What Do I Tell the Boys at the Station?

ELLIOT NUGENT (1899-1980)

1932: The Mouthpiece (w/ James Flood); Life Begins (w/ J. Flood)
1933: Three-Cornered Moon; Whistling in the Dark; If I Were Free
1934: She Loves Me Not; Strictly Dynamite; Two Alone
1935: Enter Madame; Love in Bloom; Splendor; College Scandal
1936: And So They Were Married; Wives Never Know
1937: It's All Yours
1938: Give Me a Sailor; Professor Beware
1939: Never Say Die; The Cat and the Canary
1941: Nothing but the Truth
1942: The Male Animal
1943: The Crystal Ball
1944: Up in Arms
1947: My Favorite Brunette; Welcome Stranger
1948: My Girl Tisa
1949: The Great Gatsby; Mr. Belvedere Goes to College
1950: The Skipper Surprised His Wife
1951: My Outlaw Brother
1952: Just for You

RUDOLF NUREYEV

1973: Don Quixote (w/ Robert Helpmann)

RAPHAEL NUSSBAUM

1974: Pets
1976: The Amorous Adventures of Don Quixote and Sancho Panza

CHRISTIAN NYBY (1919-)

1951: The Thing
1957: Hell on Devil's Island
1965: Young Fury; Operation CIA
1967: First to Fight

ROBERT OBER

1930: Woman Racket (w/ Albert Kelley)

ARCH OBOLER (1909-)

1945: Bewitched
1946: Strange Holiday
1947: The Arnelo Affair
1951: Five
1952: Bwana Devil
1953: The Twonky
1961: 1+1--Exploring the Kinsey Reports (Can.)
1967: The Bubble

EDMOND O'BRIEN (1915-)

1954: Shield for Murder (w/ H. W. Koch)
1961: Man Trap

JACK O'CONNELL

1963: Greenwich Village Story
1969: Revolution

JIM O'CONNOLLY

1965: The Little Ones
1968: Berserk!
1969: Valley of Gwangi
1970: Crooks and Coronets
1972: Horror on Snape Island

FRANK O'CONNOR

1930: Call of the Circus
1938: Religious Racketeers
1939: Mystic Circle Murder

WILLIAM O'CONNOR

1930: Chiselers of Hollywood
1931: Ten Nights in a Barroom; Primrose Path; Playthings of Hollywood
1932: The Drifter
1933: Her Splendid Folly
1935: Cheyenne Tornado

CLIFFORD ODETS (1903-1963)

1944: None but the Lonely Heart
1959: The Story on Page One

GEORGE O'HANLON

1959: The Rookie

MICHAEL O'HERLIHY

1966: The Fighting Prince of Donegal
1968: The One and Only Genuine Original Family Band
1969: Smith!

TOM O'HORGAN

1969: Futz
1974: Rhinoceros

DENNIS O'KEEFE (1908-1968)

1954: The Diamond Wizard (Brit.)
1955: Angela (Brit.)

DAVID O'MALLEY

1976: Guardian of the Wilderness

RON O'NEAL (1937-   )

1973: Superfly TNT

ROBERT O'NEIL

1970: The Loving Touch
1973: Wonder Women

MAURICE O'NEILL

1937: A Tenderfoot Goes West

MAX OPHULS (1902-1957) Germ.

(American films only)
1947: The Exile
1948: Letter from an Unknown Woman
1949: Caught; The Reckless Moment
1950: Vendetta (unc. w/ others)
1976: The Memory of Justice (doc.)

WYOTT ORDUNG

1956: Walk the Dark Street

RUTH ORKIN

1956: Lovers and Lollipops (w/ Morris Engel)

LESTER ORLEBECK

1940: Pioneers of the West
1941: Gauchos of Eldorado; Prairie Pioneers; Pals of the Pecos; Saddlemates; Outlaws of the Cherokee Trail; West of Cimarron
1942: Shadows on the Sage

RON ORMOND

1951: Kentucky Jubilee; Varieties on Parade; Yes Sir, Mr. Bones; King of the Bullwhip
1952: Outlaw Women (w/ S. Newfield)

ALAN ORMSBY

1974: Deranged (w/ Jeff Gillen)

KENT OSBORNE

1970: Cain's Way

JOHN O'SHAUGHNESSY

1963: The Sound of Laughter (doc.)

SAM O'STEEN

1976: Sparkle

GERD OSWALD (1916-) Germ.

1956: A Kiss Before Dying; The Brass Legend
1957: Crime of Passion; Fury at Showdown; Valeria
1958: Paris Holiday (Fr.); Screaming Mimi
1961: Brainwashed (Austria/Yugoslavia)
1966: Agent for H.A.R.M.

1969: Eighty Steps to Jonah
1971: Bunny O'Hare (alt. The Bunny O'Hare Mob)

RICHARD OSWALD (1880-1963) Austrian

(American films only)
1941: The Captain of Koepenick
1942: Isle of Missing Men
1949: The Loveable Cheat

FEDOR OZEP (1895-1949) Russian

(American films only)
1944: Three Russian Girls (w/ Henry Kesler)
1948: Whispering City

G. W. PABST (1885-1967)

(American films only)
1934: A Modern Hero

CALVIN PADGETT

1966: Secret Agent Super Dragon

ANTHONY PAGE

1968: Inadmissible Evidence
1977: I Never Promised You a Rose Garden

ALAN J. PAKULA (1928-)

1969: The Sterile Cuckoo
1971: Klute
1972: Love and Pain (and the Whole Damn Thing)
1974: The Parallax View
1976: All the President's Men
1978: Comes a Horseman
1979: Starting Over

GEORGE PAL (1908-1980) Hung.

1958: Tom Thumb
1960: The Time Machine

1961: Atlantis, the Lost Continent
1962: The Wonderful World of the Brothers Grimm (w/ H. Levin)
1964: Seven Faces of Dr. Lao

NORMAN PANAMA (1914-)

1950: The Reformer and the Redhead (w/ Melvin Frank)
1951: Strictly Dishonorable; Callaway Went Thataway (both w/ M. Frank)
1952: Above and Beyond (w/ M. Frank)
1954: Knock on Wood (w/ M. Frank)
1956: The Court Jester; That Certain Feeling (both w/ M. Frank)
1959: The Trap
1962: The Road to Hong Kong
1966: Not With My Wife, You Don't!
1969: The Maltese Bippy; How to Commit Marriage
1976: I Will, I Will ... for Now

MICHAEL J. PARADISE

1979: The Visitor

JERRY PARIS (1925-)

1968: Never a Dull Moment; Don't Raise the Bridge, Lower the River!; How Sweet It Is!
1969: Viva Max!
1970: The Grasshopper
1971: The Star-Spangled Girl

ALAN PARKER

1976: Bugsy Malone (Brit.)
1978: Midnight Express

BEN PARKER

1940: George Washington Carver (doc.)

JOE PARKER

1957: Eighteen and Anxious
1958: The Hot Angel

GORDON PARKS (1912-)

1969: The Learning Tree

1971: Shaft
1972: Superfly; Shaft's Big Score
1974: The Super Cops; Thomasine and Bushrod; Three the Hard Way
1975: Aaron Loves Angela
1976: Leadbelly

ROBERT R. PARRISH (1916-   )

1951: Cry Danger; The Mob
1952: The San Francisco Story; Assignment: Paris; My Pal Gus
1953: Shoot First (Brit.)
1955: The Purple Plain (Brit.); Lucy Gallant
1957: Fire Down Below (Brit.)
1958: Saddle the Wind
1959: The Wonderful Country
1963: In the French Style (Fr.)
1965: Up from the Beach (Fr.)
1967: Casino Royale (w/ others; Brit.); The Bobo (Brit.)
1968: Duffy (Brit.)
1969: Journey to the Far Side of the Sun (Brit.)
1971: A Town Called Hell (Span.)
1974: The Destructors (Fr.)

JAMES PARROTT (1892-1939)

1931: Pardon Us

ROBERT PARRY

1957: Tomahawk Trail

JOHN PARSONS

1974: Watched

IVAN PASSER

1971: Born to Win
1974: Law and Disorder
1976: Crime and Passion
1978: The Silver Bears

MICHAEL PATAKI

1977: Mansion of the Doomed

STUART PATON

1931: Air Police; Chinatown After Dark; First Aid; Hell Bent for Frisco; In Old Cheyenne; Is There Justice?; Mounted Fury
1935: The Silent Code; Thunderbolt
1938: Clipped Wings

(Serial)

1931: The Mystery Trooper (alt. Trail of the Royal Mounted)

ROBERT PATRICK

1970: Road to Nashville

JOHN D. PATTERSON

1975: The Legend of Earl Durand

RICHARD PATTERSON

1978: The Gentleman Tramp (doc.)

PHIL PATTON

1959: The Snow Queen (anim.)

BYRON PAUL

1966: Lt. Robin Crusoe, U.S.N.; The Tenderfoot

JOHN PAYNE (1912-)

1969: They Ran for Their Lives

LESLIE PEARCE

1930: The Fall Guy
1931: Meet the Wife

RICHARD PEARCE

1979: Heartland

SAM PECKINPAH (1925-)

1961: The Deadly Companions
1962: Ride the High Country
1965: Major Dundee
1969: The Wild Bunch
1970: The Ballad of Cable Hogue
1971: Straw Dogs
1972: The Getaway; Junior Bonner
1973: Pat Garrett and Billy the Kid
1974: Bring Me the Head of Alfredo Garcia
1975: The Killer Elite
1977: Cross of Iron (Brit./Germ.)
1978: Convoy

LARRY PEERCE

1964: One Potato, Two Potato
1966: The Big TNT Show
1967: The Incident
1969: Goodbye, Columbus
1971: The Sporting Club
1972: A Separate Peace
1973: Ash Wednesday
1975: The Other Side of the Mountain
1976: Two Minute Warning
1978: The Other Side of the Mountain, Part II
1979: The Bell Jar

BARBARA PEETERS

1970: The Dark Side of Tomorrow (w/ Jacque Beerson)
1972: Bury Me an Angel
1975: Summer School Teachers

JEAN-MARIE PELISSIE

1973: The House That Cried Murder

SCOTT PEMBROKE

1929: Should a Girl Marry?
1930: Jazz Cinderella; Last Dance; The Medicine Man
1936: The Oregon Trail
1938: Telephone Operator

ARTHUR PENN (1922-)

1958: The Left-Handed Gun

1962: The Miracle Worker
1965: Mickey One
1966: The Chase
1967: Bonnie and Clyde
1969: Alice's Restaurant
1970: Little Big Man
1973: Visions of Eight (w/ others)
1975: Night Moves
1976: The Missouri Breaks

LEO PENN

1966: A Man Called Adam

D. A. PENNEBAKER (1930-)

1968: Monterey Pop (doc.)
1972: Sweet Toronto (doc.)
1973: Keep on Rockin'

GEORGE PEPPARD (1928-)

1978: Five Days from Home

ETIENNE PERIER

1961: Bridge to the Sun
1969: The Day the Hot Line Got Hot

FRANK PERRY (1930-)

1962: David and Lisa
1963: Ladybug, Ladybug
1968: The Swimmer (w/ Sydney Pollack)
1969: Trilogy; Last Summer
1970: Diary of a Mad Housewife
1971: Doc
1972: Play It As It Lays
1974: Man on a Swing
1975: Rancho DeLuxe

PETER PERRY

1979: The Young Cycle Girls

BROOKE L. PETERS

1954: The World Dances
1957: The Unearthly
1961: Anatomy of a Psycho

DANIEL PETRIE (1920-)

1960: The Bramble Bush
1961: A Raisin in the Sun
1963: The Main Attraction; Stolen Hours
1966: The Idol; The Spy with a Cold Nose
1973: The Neptune Factor
1974: Buster and Billie
1976: Lifeguard
1977: Sybil
1978: The Betsy

BORIS PETROFF

1936: Hats Off
1952: Red Snow
1958: Outcasts of the City

JOSEPH PEVNEY (1920-)

1950: Shakedown; Undercover Girl
1951: Air Cadet; The Iron Man; The Lady from Texas; The Strange Door
1952: Meet Danny Wilson; Flesh and Fury; Just Across the Street; Because of You
1953: Desert Legion; It Happens Every Thursday; Back to God's Country
1954: Yankee Pasha; Playgirl; Three Ring Circus
1955: Six Bridges to Cross; Foxfire; Female on the Beach
1956: Away All Boats; Congo Crossing
1957: Istanbul; Tammy and the Bachelor; The Midnight Story; Man of a Thousand Faces
1958: Twilight for the Gods; Torpedo Run
1959: Cash McCall
1960: The Night of the Grizzly

JOHN PEYSER

1957: Undersea Girl
1968: The Young Warriors
1974: The Centerfold Girls

IRVING PICHEL (1891-1954)

1932: The Most Dangerous Game (w/ E. B. Schoedsack)
1933: Before Dawn
1935: She (w/ Lansing C. Holden)
1936: The Gentleman from Louisiana; Larceny on the Air
1937: Beware of Ladies; The Sheik Steps Out; The Duke Comes Back
1939: The Great Commandment
1940: Earthbound; The Man I Married; Hudson's Bay
1941: Dance Hall
1942: Secret Agent of Japan; The Pied Piper; Life Begins at Eight-Thirty
1943: The Moon Is Down; Happy Land
1944: And Now Tomorrow
1945: A Medal for Benny; Colonel Effingham's Raid
1946: Tomorrow Is Forever; The Bride Wore Boots; Temptation; O.S.S.
1947: They Won't Believe Me; Something in the Wind
1948: The Miracle of the Bells; Mr. Peabody and the Mermaid
1949: Without Honor
1950: The Great Rupert; Quicksand; Destination Moon
1951: Santa Fe
1953: Martin Luther (Brit./Germ.)
1954: Day of Triumph

CHARLES B. PIERCE

1973: The Legend of Boggy Creek (doc.)
1974: Bootleggers
1975: Winterhawk
1976: The Winds of Autumn
1977: The Town That Dreaded Sundown
1978: Grayeagle; The Norseman
1979: The Evictors

ARTHUR PIERSON

1947: Dangerous Years
1949: The Fighting O'Flynn
1951: Home Town Story

CARL PIERSON

1935: The New Frontier; Paradise Canyon; The Singing Vagabond

FRANK PIERSON

1976: A Star Is Born
1978: King of the Gypsies

DAVID PINCUS

1934: These Thirty Years

ED PINCUS

1978: Life and Other Anxieties (doc. w/ Steven Ascher)

PHILLIP PINE

1972: The Cat Ate the Canary
1975: Pot! Parents! Police!

WILLIAM H. PINE (1896-)

1942: Aerial Gunner
1945: Swamp Fire
1947: Seven Were Saved
1949: Dynamite

SIDNEY PINK (1916-)

1961: Journey to the Seventh Planet
1962: Reptilicus (Den.)
1965: Finger on the Trigger
1966: The Tall Woman

ERNEST PINTOFF (1931-)

1965: Harvey Middleman, Fireman
1971: Who Killed Mary What's'ername?
1972: Dynamite Chicken (doc.)
1979: Jaguar Lives!

ROBERT PIROSH (1910-)

1951: Go for Broke!
1952: Washington Story
1954: Valley of the Kings
1955: The Girl Rush
1957: Spring Reunion

H. LEE POGOSTIN

1969: Hard Contract

SIDNEY POITIER (1924-)

1972: Buck and the Preacher
1973: A Warm December
1974: Uptown Saturday Night
1975: Let's Do It Again
1977: A Piece of the Action

JAMES POLAKOFF

1975: Sunburst
1979: Love and the Midnight Auto Supply

ROMAN POLANSKI (1933-) Pol.

(American films only)
1967: The Vampire Killers
1968: Rosemary's Baby
1971: Macbeth
1974: Chinatown

BARRY POLLACK

1972: Cool Breeze
1973: This Is a Hijack

SYDNEY POLLACK (1934-)

1965: The Slender Thread
1966: This Property Is Condemned
1968: The Scalphunters; The Swimmer (unc. w/ F. Perry)
1969: Castle Keep; They Shoot Horses, Don't They?
1972: Jeremiah Johnson
1973: The Way We Were
1975: Three Days of the Condor; The Yakuza
1977: Bobby Deerfield
1979: The Electric Horseman

BUD POLLARD

1930: Danger Man
1931: Alice in Wonderland; Rio's Road to Hell; Voice of the Jungle
1932: Black King
1933: Victims of Persecution
1937: The Death March
1946: Beware; Tall, Tan and Terrific
1947: It Pays to Be Funny; Road to Hollywood

HARRY A. POLLARD (1883-1934)

1929: Show Boat; Tonight at Twelve
1930: Undertow
1931: Shipmates; The Prodigal
1932: Fast Life; When a Feller Needs a Friend

JACK POLLEXFEN

1954: Dragon's Gold (w/ Aubrey Wisberg)
1956: The Indestructible Man

ABRAHAM POLONSKY (1910-   )

1948: Force of Evil
1969: Tell Them Willie Boy Is Here
1971: Romance of a Horsethief

ROY J. POMEROY

1928: Interference (w/ L. Mendes)
1930: Inside the Lines
1934: Shock

LEO C. POPKIN

1939: One Dark Night; Reform School
1951: The Well (w/ Russell Rouse)

CHARLES A. POST

1932: Single-Handed Sanders

TED POST (1918-   )

1956: The Peacemaker
1959: The Legend of Tom Dooley
1968: Hang 'Em High
1970: Beneath the Planet of the Apes
1973: The Baby; The Harrad Experiment; Magnum Force
1975: Whiffs
1978: Go Tell the Spartans
1979: Good Guys Wear Black

H. C. (HENRY) POTTER (1904-1977)

1936: Beloved Enemy

1937: Wings over Honolulu
1938: Romance in the Dark; Shopworn Angel; The Cowboy and the Lady
1939: The Story of Vernon and Irene Castle; Blackmail
1940: Congo Maisie; Second Chorus
1941: Hellzapoppin
1943: Mr. Lucky
1947: The Farmer's Daughter; A Likely Story
1948: Mr. Blandings Builds His Dream House; The Time of Your Life; You Gotta Stay Happy
1950: The Miniver Story
1955: Three for the Show
1957: Top Secret Affair

DICK POWELL (1904-1963)

1953: Split Second
1956: The Conqueror; You Can't Run Away from It
1957: The Enemy Below
1958: The Hunters

STANLEY PRAGER

1968: The Bang Bang Kid (Span.)
1970: Madigan's Millions

GILBERT PRATT

1934: Elmer and Elsie

MICHAEL PREECE

1979: The Prize Fighter

OTTO PREMINGER (1906-)

1936: Under Your Spell
1937: Danger-Love at Work
1943: Margin for Error
1944: In the Meantime, Darling; Laura
1945: Fallen Angel; A Royal Scandal (w/ Ernst Lubitsch)
1946: Centennial Summer
1947: Daisy Kenyon; Forever Amber
1949: The Fan; Whirlpool
1950: Where the Sidewalk Ends
1951: The 13th Letter
1953: Angel Face; The Moon Is Blue
1954: River of No Return; Carmen Jones

1955: The Court-Martial of Billy Mitchell; The Man with the Golden Arm
1957: St. Joan
1958: Bonjour Tristesse
1959: Anatomy of a Murder; Porgy and Bess
1960: Exodus
1962: Advise and Consent
1963: The Cardinal
1965: Bunny Lake Is Missing; In Harm's Way
1967: Hurry Sundown
1969: Skidoo!
1970: Tell Me That You Love Me, Junie Moon
1971: Such Good Friends
1975: Rosebud
1979: The Human Factor

CHRIS PRENTISS

1976: Goin' Home

FRED PRESSBURGER

1956: Crowded Paradise

MICHAEL PRESSMAN

1976: The Great Texas Dynamite Chase
1977: The Bad News Bears in Breaking Training
1979: Boulevard Nights

WILL PRICE

1949: Strange Bargain
1950: Tripoli

HAROLD PRINCE

1970: Something for Everyone
1978: A Little Night Music

LEROY PRINZ (1895-)

1941: All-American Co-ed; Fiesta

FRANCO PROSPERI

1975: Ripped-Off

MOHY QUANDOUR

1974: The Spectre of Edgar Allan Poe

THOMAS QUILLEN

1975: Pursuit

RICHARD QUINE (1920-)

1948: Leather Gloves (w/ W. Asher)
1951: Sunny Side of the Street; Purple Heart Diary
1952: Rainbow 'Round My Shoulder; Sound Off
1953: All Ashore; Siren of Bagdad; Cruisin' Down the River
1954: Drive a Crooked Road; Pushover; So This Is Paris
1955: My Sister Eileen
1956: The Solid Gold Cadillac
1957: Full of Life; Operation Mad Ball
1958: Bell, Book and Candle
1959: It Happened to Jane
1960: Strangers When We Meet; World of Suzie Wong
1962: The Notorious Landlady
1964: Paris When It Sizzles; Sex and the Single Girl
1965: How to Murder Your Wife; Synanon
1967: Hotel; Oh, Dad, Poor Dad, Mama's Hung You in the Closet and I'm Feeling so Sad
1969: A Talent for Loving
1970: The Moonshine War
1974: W
1979: The Prisoner of Zenda

ANTHONY QUINN (1915-)

1958: The Buccaneer

JOSE QUINTERO

1961: The Roman Spring of Mrs. Stone

AL RABOCH

1934: Rocky Rhodes
1935: The Crimson Trail

MICHAEL RAE

1978: Laserblast

BOB RAFELSON (1934-)

1968: Head
1970: Five Easy Pieces
1972: The King of Marvin Gardens
1976: Stay Hungry

STEWART RAFFILL

1971: The Tender Warrior
1975: The Adventures of the Wilderness Family
1976: Across the Great Divide
1978: The Sea Gypsies

ALAN RAFKIN

1965: Ski Party
1966: The Ghost and Mr. Chicken
1967: The Ride to Hangman's Tree
1968: Nobody's Perfect; The Shakiest Gun in the West
1969: Angel in My Pocket
1971: How to Frame a Figg

ED RAGOZZINI

1978: Sasquatch

JOEL RAPP

1959: High School Big Shot
1960: Battle of Blood Island

PAUL RAPP

1976: Go for It (doc.)

MARK RAPPAPORT

1978: Scenic Route

IRVING RAPPER (1898-)

1941: Shining Victory; One Foot in Heaven
1942: The Gay Sisters; Now, Voyager
1944: The Adventures of Mark Twain
1945: The Corn Is Green; Rhapsody in Blue
1946: Deception

1947: The Voice of the Turtle
1949: Anna Lucasta
1950: The Glass Menagerie
1952: Another Man's Poison (Brit.)
1953: Forever Female; Bad for Each Other
1956: Strange Intruder; The Brave One
1958: Marjorie Morningstar
1959: The Miracle
1962: Joseph and His Brethren
1970: The Christine Jorgensen Story
1978: Born Again

STEVE RASH

1978: The Buddy Holly Story

HARRY RASKY

1979: Arthur Miller on Home Ground (doc.)

GREGORY RATOFF (1897-1961)

1936: Sins of Man (w/ Otto Brower)
1937: The Lancer Spy
1939: Wife, Husband and Friend; Rose of Washington Square; Hotel for Women; Intermezzo; Barricade; Daytime Wife
1940: I Was an Adventuress; Public Deb Number One
1941: Adam Had Four Sons; The Men in Her Life; The Corsican Brothers
1942: Two Yanks in Trinidad; Footlight Serenade
1943: Something to Shout About; The Heat's On; Song of Russia
1944: Irish Eyes Are Smiling
1945: Where Do We Go from Here?; Paris Underground
1946: Do You Love Me?
1947: Moss Rose
1949: Black Magic
1950: If This Be Sin (Brit.)
1951: Operation X (Brit.)
1953: Taxi
1956: Abdullah's Harem (Egypt)
1960: Oscar Wilde (Brit.)

JOHN RAWLINS (1902-   )

1938: State Police; Young Fugitives; The Missing Guest; Air Devils
1940: The Leatherpushers
1941: Six Lessons from Madame La Zonga; A Dangerous Game; Mr. Dynamite; Mutiny in the Arctic; Men of the Timberland; Raiders of the Desert

1942: Bombay Clipper; Unseen Enemy; Mississippi Gambler; Half Way to Shanghai; Sherlock Holmes and the Voice of Terror; The Great Impersonation; Arabian Nights; Torpedo Boat
1943: We've Never Been Licked
1944: Ladies Courageous
1945: Sudan
1946: Strange Conquest; Her Adventurous Night
1947: Dick Tracy's Dilemma; Dick Tracy Meets Gruesome
1948: The Arizona Ranger; Michael O'Halloran
1949: Massacre River
1950: Boy from Indiana; Rogue River
1951: Fort Defiance
1953: Shark River
1958: Lost Lagoon

(Serials)

1940: Junior G-Men; The Green Hornet Strikes Again (both w/ Ford Beebe)
1941: Sea Raiders (w/ F. Beebe)
1942: Overland Mail (w/ F. Beebe)

ALBERT RAY

1929: Molly and Me
1930: Call of the West; Her Unborn Child (w/ Charles McGrath); Kathleen Mavourneen
1932: The Thirteenth Guest; Unholy Love; Guilty or Not Guilty
1933: West of Singapore; A Shriek in the Night; The Intruder
1934: Dancing Man
1935: St. Louis Woman; Marriage Bargain
1936: Everyman's Law; Lawless Land; Undercover Man
1939: Desperate Trails

BERNARD B. RAY

1934: Rawhide Mail; Mystery Ranch; Loser's End
1935: Fast Bullets; Midnight Phantom; Now or Never; The Silver Bullet; Coyote Trails; Rio Rattler; Silent Valley; Texas Jack; Never Too Late
1936: Ambush Valley; Caryl of the Mountains; Prince of Rustlers; Riding On; Millionaire Kid; Roamin' Wild; The Speed Reporter; The Test; Santa Fe Trail; Trigger Tom; Step On It; Vengeance of Rannah
1937: The Silver Trail
1938: It's All in Your Mind; Santa Fe Rider
1939: Smoky Trails
1940: Broken Strings
1941: Dangerous Lady; Law of the Timber
1942: Too Many Women; House of Errors
1946: Buffalo Bill Rides Again

1947: Hollywood Barn Dance
1950: Timber Fury
1952: Buffalo Bill in Tomahawk Territory
1953: Hollywood Thrill Makers
1960: Spring Affair

NICHOLAS RAY (1911-1979)

1947: Your Red Wagon
1948: The Twisted Road (ret. They Live by Night)
1949: Knock on Any Door; A Woman's Secret
1950: Born to Be Sad; In a Lonely Place
1951: Flying Leathernecks; On Dangerous Ground (w/ Ida Lupino); This Man Is Mine
1952: The Lusty Men
1954: Johnny Guitar
1955: Rebel Without a Cause; Run for Cover
1956: Bigger Than Life; Hot Blood
1957: The True Story of Jesse James
1958: Bitter Victory (Fr.); Party Girl; Wind Across the Everglades
1959: The Savage Innocents
1961: King of Kings
1963: 55 Days at Peking

HERMAN C. RAYMAKER

1932: Trailing the Killer
1934: Adventure Girl

GENE RAYMOND (1908-)

1948: Million Dollar Weekend

BILL REBANE

1975: The Giant Spider Invasion
1979: The Capture of Bigfoot

CAROL REED (1906-1976) Brit.

(American films only)
1956: Trapeze
1965: The Agony and the Ecstasy
1970: Flap

JAMES REED

1973: Tarzana, the Wild Girl

JAY THEODORE REED (1887-1959)

1936: Lady, Be Careful
1937: Double or Nothing
1938: Tropic Holiday
1939: I'm from Missouri; What a Life!
1940: Those Were the Days
1941: Life with Henry; Her First Beau

JOEL M. REED

1972: Career Bed
1975: Dragon Lady

LUTHER REED (1888-1961)

1929: Rio Rita
1930: Hit the Deck; Dixiana
1935: Convention Girl

ROLAND REED

1936: In Paris A.W.O.L.
1937: Red Lights Ahead

JULIA REICHERT

1977: Union Maids (doc.)

THOMAS REICHMAN

1968: Mingus (doc.)

DOROTHY REID (MRS. WALLACE REID) (1895-1977)

1929: Linda
1933: Sucker Money (w/ Melville Shyer)
1934: The Road to Ruin (w/ Melville Shyer); A Woman Condemned

HARALD REIGL

1974: Chariots of the Gods (doc.)

CARL REINER (1922-)

1967: Enter Laughing

1969: The Comic
1970: Where's Poppa?
1977: Oh, God!
1978: The One and Only
1979: The Jerk

GOTTFRIED REINHARDT (1911-   ) Germ.

(American films only)
1952: Invitation
1953: Story of Three Loves (w/ V. Minnelli)
1954: Betrayed
1961: Town Without Pity
1965: Situation Hopeless--But Not Serious

JOHN REINHARDT

1936: Captain Calamity
1947: The Guilty; High Tide; For You I Die; Ambush
1948: Sofia; Open Secret
1951: Chicago Calling

MAX REINHARDT (1873-1943) Austrian

(American films only)
1935: A Midsummer Night's Dream (w/ William Dieterle)

IRVING REIS (1906-1953)

1940: One Crowded Night; I'm Still Alive
1941: Footlight Fever; The Gay Falcon; Weekend for Three; A Date with the Falcon
1942: The Falcon Takes Over; The Big Street
1946: The Crack-Up
1947: The Bachelor and the Bobby-Soxer
1948: All My Sons; Enchantment
1949: Roseanna McCoy; Dancing in the Dark
1950: Three Husbands; Of Men and Music (w/ Alex Hammid)
1951: New Mexico
1952: The Four Poster

WALTER REISCH (1903-   ) Austrian

(American films only)
1947: Song of Scheherazade

ALLEN REISNER

1957: All Mine to Give
1958: St. Louis Blues

CHARLES F. REISNER (see CHARLES F. RIESNER)

KAREL REISZ (1926-) Czech.

(American films only)
1974: The Gambler
1978: Who'll Stop the Rain

WOLFGANG REITHERMAN

1973: Robin Hood (anim.)

IVAN REITMAN

1973: Cannibal Girls
1979: Meatballs

JEAN RENOIR (1894-1980) Fr.

(American films only)
1941: Swamp Water
1943: This Land Is Mine
1944: Salute to France (doc.)
1945: The Southerner
1946: The Diary of a Chambermaid
1947: The Woman on the Beach
1951: The River (Ind.)

HARRY J. REVIER (1889-)

1930: The Convict's Code
1935: The Lost City (formerly serial)
1936: The Lash of the Penitentes
1938: Child Bride

BURT REYNOLDS (1936-)

1976: Gator
1978: The End

CLARK E. REYNOLDS

1958: Gunman from Laredo

QUENTIN REYNOLDS

1950: Death of a Dream (doc.)
1960: Justice and Caryl Chessman (doc.)

SHELDON REYNOLDS

1956: Foreign Intrigue
1969: Assignment to Kill

WILLIAM "RED" REYNOLDS

1960: Chartroose Caboose

RON RICE

1962: The Flower Thief

DAVID LOWELL RICH (c. 1923-)

1957: No Time to Be Young
1958: Senior Prom
1959: Hey Boy! Hey Girl!; Have Rocket, Will Travel
1966: Madame X; The Plainsman
1967: Rosie!
1968: Three Guns for Texas (w/ Paul Stanley, Earl Bellamy); A Lovely Way to Die
1969: Eye of the Cat
1973: That Man Bolt (w/ Henry Levin; U.S./Hong Kong)

JOHN RICH (1925-)

1963: Wives and Lovers
1964: The New Interns; Roustabout
1965: Boeing Boeing
1967: Easy Come, Easy Go

DICK RICHARDS

1972: The Culpepper Cattle Company
1975: Rafferty and the Gold Dust Twins; Farewell, My Lovely (Brit.)
1977: March or Die

TONY RICHARDSON (1928-) Brit.

(American films only)
1961: Sanctuary
1965: The Loved One
1973: A Delicate Balance

WILLIAM RICHERT

1973: First Position (doc.)
1939: Winter Kills

RICHARD RICHTER

1979: Vietnam: An American Journey (doc.)

CHARLES F. RIESNER (also CHARLES F. REISNER) (1887-1962)

1928: Brotherly Love
1929: Noisy Neighbors; China Bound; The Hollywood Revue of 1929; Road Show
1930: Chasing Rainbows; Caught Short; Love in the Rough
1931: Reducing; Stepping Out; Politics; Flying High
1932: Divorce in the Family
1933: The Chief
1934: You Can't Buy Everything; The Show-Off; Student Tour
1935: The Winning Ticket; It's in the Air
1936: Everybody Dance (Brit.)
1937: Murder Goes to College; Sophie Lang Goes West; Manhattan Merry-Go-Round
1939: Winter Carnival
1941: The Big Store
1942: This Time for Keeps
1943: Harrigan's Kid
1944: Meet the People; Lost in a Harem
1948: The Cobra Strikes; In This Corner
1950: The Traveling Saleswoman; L'ultima cena (Ital.)

DEAN RIESNER

1947: Bill and Coo

ARTHUR RIPLEY (1895-1961)

1938: I Met My Love Again (w/ Joshua Logan)
1942: Prisoner of Japan
1944: Voice in the Wind
1946: The Chase
1958: Thunder Road

ROBERT RISKIN (1897-1955)

1937: When You're in Love

MICHAEL RITCHIE (1938-)

1969: Downhill Racer
1972: The Candidate; Prime Cut
1975: Smile
1976: The Bad News Bears
1977: Semi-Tough
1979: An Almost Perfect Affair

MARTIN RITT (1919-)

1957: Edge of the City; No Down Payment
1958: The Long, Hot Summer
1959: The Black Orchid; The Sound and the Fury
1960: Five Branded Women (Ital.)
1961: Paris Blues
1962: Hemingway's Adventures of a Young Man
1963: Hud
1964: The Outrage
1965: The Spy Who Came in from the Cold
1967: Hombre
1968: The Brotherhood
1970: The Molly Maguires; The Great White Hope
1972: Sounder; Pete 'n' Tillie
1974: Conrack
1976: The Front
1978: Casey's Shadow
1979: Norma Rae

LLOYD RITTER

1956: Secrets of the Reef (w/ others)

JOAN RIVERS

1978: Rabbit Test

HAL ROACH (1892-)

1930: Man of the North
1933: The Devil's Brother (w/ Charles Rogers)
1939: Captain Fury; The Housekeeper's Daughter
1940: One Million B.C. (w/ Hal Roach, Jr.); Turnabout
1941: Road Show (w/ Gordon Douglas, H. Roach, Jr.)

HAL ROACH, JR. (1921-1972)

1940: One Million B.C. (w/ Hal Roach, Sr.)

1941: Road Show (w/ Gordon Douglas, H. Roach, Sr.)
1942: Dudes Are Pretty People
1943: Calaboose; Prairie Chickens

SEYMOUR ROBBIE

1970: C. C. and Company
1974: Marco

JEROME ROBBINS (1918-)

1961: West Side Story (w/ Robert Wise)

MATTHEW ROBBINS

1978: Corvette Summer

ARTHUR ROBERSON

1974: Black Hooker

ALAN ROBERTS

1970: The Zodiac Couples (w/ Bob Stein)
1975: Panorama Blue

BOB ROBERTS

1971: Sweet Saviour

CHARLES E. ROBERTS

1933: Flaming Signal (w/ G. Jeske); Corruption
1935: Adventurous Knights
1941: Hurry, Charlie, Hurry

FLORIAN ROBERTS (see ROBERT FLOREY)

STEPHEN ROBERTS (1895-1936)

1932: Sky Bride; Lady and Gent; The Night of June 13th; If I Had a Million (w/ others)
1933: The Story of Temple Drake; One Sunday Afternoon
1934: The Trumpet Blows; Romance in Manhattan

1935: Star of Midnight; The Man Who Broke the Bank at Monte Carlo
1936: The Lady Consents; The Ex-Mrs. Bradford

CLIFF ROBERTSON (1925-)

1972: J. W. Coop

HUGH A. ROBERTSON

1972: Melinda

JOHN S. ROBERTSON (1878-1964)

1929: Shanghai Lady
1930: Night Ride; Captain of the Guard (w/ P. Fejos); Madonna of the Streets
1931: Beyond Victory; The Phantom of Paris
1932: Little Orphan Annie
1933: One Man's Journey
1934: The Crime Doctor; His Greatest Gamble; Wednesday's Child
1935: Grand Old Girl; Our Little Girl; Captain Hurricane

CASEY ROBINSON (1903-)

1932: Renegades of the West

CHRIS ROBINSON

1979: Sunshine Run

DICK ROBINSON

1975: Birth of a Legend (doc.); Brother of the Wind; Poor Pretty Eddie
1979: Redneck Country

R. D. ROBINSON

1971: The World Is Just a "B" Movie

MARK ROBSON (1913-)

1943: The Seventh Victim; The Ghost Ship
1944: Youth Runs Wild

1945: Isle of the Dead
1946: Bedlam
1949: Champion; Home of the Brave; Roughshod; My Foolish Heart
1950: Edge of Doom
1951: Bright Victory; I Want You
1953: Return to Paradise
1954: Hell Below Zero (Brit.); Phffft; The Bridges at Toko-Ri
1955: A Prize of Gold; Trial
1956: The Harder They Fall
1957: The Little Hut; Peyton Place
1958: The Inn of the Sixth Happiness
1960: From the Terrace
1963: Nine Hours to Rama; The Prize
1965: Von Ryan's Express
1966: Lost Command
1967: Valley of the Dolls
1969: Daddy's Gone A-Hunting
1971: Happy Birthday, Wanda June
1972: Limbo
1974: Earthquake
1979: Avalanche Express

PAT ROCCO

1975: Drifter

FRANC RODDAM

1979: Quadrophenia

ISMAEL RODRIGUEZ

1956: Daniel Boone, Trail Blazer (w/ A. C. Gannaway); The Beast of Hollow Mountain (w/ E. Nassour)

MICHAEL ROEMER (1928-)

1964: Nothing But a Man

ALBERT S. ROGELL (1901-)

1929: Phantom City; Cheyenne; Lone Wolf's Daughter; California Mail; Flying Marine
1930: Painted Faces; Mamba
1931: Aloha; Sweepstakes; The Tip-Off; The Suicide Fleet
1932: Carnival Boat; The Rider of Death Valley
1933: Air Hostess; Below the Sea; The Wrecker; East of Fifth Avenue

1934: Fog; No More Women; The Hell Cat; Among the Missing; Name the Woman; Fugitive Lady
1935: Unknown Woman; Air Hawks; Atlantic Adventure; Escape from Devil's Island (alt. Song of the Damned)
1936: Roaming Lady; Grand Jury
1937: Murder in Greenwich Village
1938: Start Cheering; The Lone Wolf in Paris; City Streets; The Last Warning
1939: For Love or Money; Hawaiian Nights; Laugh It Off
1940: Private Affairs; I Can't Give You Anything but Love, Baby; Argentine Nights; Li'l Abner
1941: The Black Cat; Tight Shoes; Public Enemies
1942: Jail House Blues; Sleepytime Gal; True to the Army; Butch Minds the Baby; Priorities on Parade; Youth on Parade
1943: Hit Parade of 1943; In Old Oklahoma
1945: Love, Honor, and Goodbye
1946: Earl Carroll Sketchbook; The Magnificent Rogue
1947: Heaven Only Knows
1948: Northwest Stampede
1949: The Song of India
1950: The Admiral Was a Lady
1956: Shadow of Fear (Brit.)

CHARLES R. ROGERS

1933: The Devil's Brother (orig. Fra Diavalo; w/ Hal Roach)
1934: Babes in Toyland (w/ Gus Meins)
1936: The Bohemian Girl (w/ James W. Horne)

LIONEL ROGOSIN (1924-)

1957: On the Bowery (doc.)
1960: Come Back, Africa (doc.)
1970: Black Roots (doc.)
1972: Black Fantasy (doc.)
1973: Woodcutters of the Deep South (doc.)

GEORGE ROLAND

1932: Joseph in the Land of Egypt

SUTTON ROLEY

1972: The Loners
1974: Chosen Survivors

EDDIE ROMERO

1963: The Raiders of Leyte Gulf (Philippines); Cavalry Command

1964: The Walls of Hell (w/ Gerardo DeLeon); Moro Witch Doctor
1965: The Ravagers
1968: Brides of Blood (w/ G. DeLeon)
1969: Blood Demon; Mad Doctor of Blood Island
1970: Beast of the Yellow Night (Philippines)
1971: Beast of Blood
1972: Twilight People
1973: Black Mama, White Mama
1974: Beyond Atlantis (Philippines); Savage Sisters
1975: The Woman Hunt

GEORGE A. ROMERO (c. 1939-)

1968: Night of the Living Dead
1972: There's Always Vanilla (alt. The Affair)
1973: The Crazies; Jack's Wife (alt. Hungry Wives)
1979: Dawn of the Dead

CHARLES R. RONDEAU

1958: The Littlest Hobo; Devil's Partner
1960: The Threat
1975: Trainride to Hollywood

CONRAD ROOKS

1967: Chappaqua

MICKEY ROONEY (1922-)

1951: My True Story
1960: The Private Lives of Adam and Eve (w/ A. Zugsmith)

ANDRE ROOSEVELT

1938: Man Hunters of the Caribbean (doc. w/ E. Scott)

WELLS ROOT

1936: The Bold Caballero
1942: Mokey

CLIFF ROQUEMORE

1976: The Human Tornado

SHERMAN A. ROSE

1955: Target Earth
1956: Magnificent Roughnecks
1958: Tank Battalion

WILLIAM L. ROSE

1968: Pamela, Pamela, You Are...

PHIL ROSEN (1888-1951)

1929: The Faker; Peacock Fan; The Phantom in the House
1930: The Rampant Age; The Lotus Lady; Worldly Goods; Extravagance
1931: Second Honeymoon; The Two-Gun Man; Alias the Bad Man; The Arizona Terror; Range Law; Branded Men; The Pocatello Kid
1932: The Gay Buckaroo; The Texas Gun Fighter; Whistlin' Dan; Lena Rivers; The Vanishing Frontier; Klondike; A Man's Land
1933: Young Blood; Self Defense; The Phantom Broadcast; The Sphinx; Black Beauty; Devil's Mate; Hold the Press
1934: Beggars in Ermine; Shadows of Sing Sing; Picture Brides; Cheaters; Take the Stand; Dangerous Corner; Woman in the Dark; Little Men; West of the Pecos
1935: Death Flies East; Unwelcome Stranger; Born to Gamble
1936: The Calling of Dan Matthews; Tango; The Bridge of Sighs; Three of a Kind; Easy Money; It Couldn't Have Happened; Brilliant Marriage; The President's Mystery; Missing Girls; Ellis Island
1937: Two Wise Maids; Jim Hanvey, Detective; It Could Happen to You; Roaring Timber; Youth on Parole
1938: The Marines Are Here
1939: Ex-Champ; Missing Evidence
1940: Double Alibi; Forgotten Girls; The Crooked Road; Phantom of Chinatown; Queen of the Yukon
1941: The Roar of the Press; Paper Bullets (alt. Crime, Inc.; Gangs, Inc.); Murder by Invitation; The Deadly Game; Spooks Run Wild; I Killed That Man
1942: Road to Happiness; The Man With Two Lives; The Mystery of Marie Roget
1943: You Can't Beat the Law; A Gentle Gangster; Wings over the Pacific
1944: Charlie Chan in the Secret Service; The Chinese Cat; Return of the Ape Man; Black Magic; Call of the Jungle; The Jade Mask; Army Wives
1945: The Scarlet Clue; The Red Dragon; Captain Tugboat Annie; In Old New Mexico
1946: The Shadow Returns; The Strange Mr. Gregory; Step by Step
1949: The Secret of St. Ives

STUART ROSENBERG (1928-)

1960: Murder, Inc. (w/ Burt Balaban)
1961: Question 7 (Germ.)
1967: Cool Hand Luke
1969: The April Fools; The Lenny Bruce Story
1970: Move; WUSA
1972: Pocket Money
1973: The Laughing Policeman
1976: The Drowning Pool
1979: The Amityville Horror; Love and Bullets

DICK ROSS

1971: The Late Liz

FRANK ROSS (1904-)

1951: The Lady Says No

HERBERT ROSS (1927-)

1969: Goodbye, Mr. Chips
1970: The Owl and the Pussycat; T. R. Baskin
1972: Play It Again, Sam
1973: The Last of Sheila
1975: The Sunshine Boys; Funny Lady
1976: The Seven Percent Solution
1977: The Goodbye Girl; The Turning Point
1978: California Suite
1979: Nijinsky

NAT ROSS

1929: College Love

ROBERT ROSSEN (1908-1966)

1947: Johnny O'Clock; Body and Soul
1949: All the King's Men
1951: The Brave Bulls
1955: Mambo (Ital.)
1956: Alexander the Great
1957: Island in the Sun
1959: They Came to Cordura
1961: The Hustler
1964: Lilith

ARTHUR ROSSON (1889-1960)

1929: The Winged Horseman (w/ B. R. Eason); Points West; The Long, Long Trail
1930: The Mounted Stranger; Trailing Trouble; The Concentratin' Kid
1933: Hidden Gold; Flaming Guns
1937: Boots of Destiny; Trailin' Trouble

RICHARD ROSSON (1894-1953)

1929: The Very Idea (w/ W. LeBaron)
1935: West Point of the Air
1937: Behind the Headlines; Hideaway
1943: Corvette K-225

TED ROTER

1979: One Page of Love

CY ROTH

1953: Combat Squad
1955: Air Strike
1956: Fire Maidens of Outer Space

MURRAY ROTH

1933: Don't Bet on Love
1934: Million Dollar Ransom; Harold Teen
1935: Chinatown Squad
1936: Flying Hostess

JOSEPH ROTHMAN

1938: Dynamite Delaney

STEPHANIE ROTHMAN

1966: Blood Bath (w/ Jack Hill)
1967: It's a Bikini World
1970: The Student Nurses
1971: The Velvet Vampire
1972: Group Marriage
1973: Terminal Island
1974: The Working Girls

HERMAN ROTSTEN

1944: The Unwritten Code

RUSSELL ROUSE (c. 1916-)

1951: The Well (w/ Leo Popkin)
1952: The Thief
1953: Wicked Woman
1955: New York Confidential
1956: The Fastest Gun Alive
1957: House of Numbers
1959: Thunder in the Sun
1964: A House Is Not a Home
1966: The Oscar
1967: The Caper of the Golden Bulls

ROY ROWLAND (c. 1910-)

1943: A Stranger in Town; Lost Angel
1945: A Song for Miss Julie; Our Vines Have Tender Grapes
1946: Boys' Ranch
1947: The Romance of Rosy Ridge; Killer McCoy
1948: Tenth Avenue Angel
1949: Scene of the Crime
1950: The Outriders; Two Weeks With Love
1951: Excuse My Dust
1952: Bugles in the Afternoon
1953: Affair with a Stranger; The 5,000 Fingers of Dr. T; The Moonlighter
1954: Witness to Murder; Rogue Cop
1955: Hit the Deck; Many Rivers to Cross
1956: Meet Me in Las Vegas; These Wilder Years; Slander
1957: Gun Glory
1958: The Seven Hills of Rome (Ital.)
1963: The Girl Hunters
1965: Gunfighters of Casa Grande
1967: The Sea Pirate

WILLIAM ROWLAND

1939: Perfida
1946: Flight to Nowhere
1948: Woman in the Night
1970: The Wild Scene

J. WALTER RUBEN (1899-1942)

1931: The Public Defender; Secret Service

1932: The Phantom of Crestwood; Roadhouse Murder
1933: The Great Jasper; No Marriage Ties; Aces of Aces; No Other Woman
1934: Man of Two Worlds; Success at Any Price; Where Sinners Meet
1935: Public Hero Number One; Riff Raff; Java Head
1936: Old Hutch; Trouble for Two
1937: Good Old Soak
1938: Bad Man of Brimstone

PERCIVAL RUBENS

1969: Strangers at Sunrise
1973: Mr. Kingstreet's War

B. ROBERT RUBIN

1946: Hollywood Hi

ALAN RUDOLPH

1972: Premonition
1976: Barn of the Naked Dead
1977: Welcome to L.A.
1978: Remember My Name

OSCAR RUDOLPH

1954: The Rocket Man
1961: Twist Around the Clock
1962: Don't Knock the Twist; The Wild Westerners

WESLEY RUGGLES (1889-1972)

1929: Scandal; Street Girl; Girl Overboard; Condemned
1930: Honey; The Sea Bat
1931: Cimarron; Are These Our Children?
1932: Roar of the Dragon; No Man of Her Own
1933: The Monkey's Paw; College Humor; I'm No Angel
1934: Bolero; Shoot the Works
1935: The Gilded Lily; Accent on Youth; The Bride Comes Home
1936: Valiant Is the Word for Carrie
1937: I Met Him in Paris; True Confession
1938: Sing You Sinners
1939: Invitation to Happiness
1940: Too Many Husbands; Arizona
1941: You Belong to Me
1942: Somewhere I'll Find You

1943: Slightly Dangerous
1944: See Here, Private Hargrove
1953: My Heart Goes Crazy (Brit.; alt. London Town)

RICHARD RUSH (1930-)

1960: Too Soon to Love
1963: Of Love and Desire
1967: Thunder Alley; Hell's Angels on Wheels; The Fickle Finger of Fate; A Man Called Dagger
1968: Psych-Out; The Savage Seven
1970: Getting Straight
1974: Freebie and the Bean
1978: The Stuntman

RUSTY RUSSELL

1973: The New York Experience

WILLIAM D. RUSSELL (1908-1968)

1946: Our Hearts Were Growing Up
1947: Ladies' Man; Dear Ruth
1948: The Sainted Sisters
1949: The Green Promise; Bride for Sale
1951: Best of the Badmen

FRANK RYAN (1907-1947)

1942: Call Out the Marines (w/ W. Hamilton)
1943: Hers to Hold
1944: Can't Help Singing
1945: Patrick the Great
1946: So Goes My Love

MARK RYDELL (1934-)

1968: The Fox
1969: The Reivers
1972: The Cowboys
1973: Cinderella Liberty
1976: Harry and Walter Go to New York
1979: The Rose

WILLIAM SACHS

1976: Secrets of the Gods
1978: The Incredible Melting Man
1979: Van Nuys Blvd.

WILLIAM A. SACKHEIM

1945: Let's Go Steady

EDDIE SAETA

1973: Doctor Death, Seeker of Souls

BORIS SAGAL (1923-1981)

1963: Dime with a Halo; Twilight of Honor
1965: Girl Happy
1966: Made in Paris
1969: The Thousand Plane Raid
1970: Mosquito Squadron (Brit.)
1971: The Omega Man

MALCOLM ST. CLAIR (1897-1952)

1929: Side Street; Night Parade; The Canary Murder Case
1930: Montana Moon; Dangerous Nan McGrew; The Boudoir Diplomat; Remote Control (w/ N. Grinde)
1933: Goldie Gets Along; Olsen's Big Moment
1937: Crack-Up; Time Out for Romance; She Had to Eat; Born Reckless; Dangerously Yours
1938: A Trip to Paris; Safety in Numbers; Down on the Farm; Everybody's Baby
1939: Quick Millions; The Jones Family in Hollywood
1940: Young As You Feel; Meet the Missus
1941: The Bashful Bachelor
1942: Over My Dead Body; The Man in the Trunk
1943: The Dancing Masters; Jitterbugs; Two Weeks to Live; Swing Out the Blues
1944: The Big Noise
1945: The Bullfighters
1948: Arthur Takes Over; Fighting Back

RAYMOND ST. JACQUES (1930-)

1973: Book of Numbers

CHRISTOPHER ST. JOHN

1972: Top of the Heap

GENE SAKS (1921-)

1967: Barefoot in the Park
1968: The Odd Couple
1969: Cactus Flower
1972: Last of the Red Hot Lovers
1974: Mame

RICHARD SALE (1911-)

1947: Spoilers of the North
1948: Campus Honeymoon
1950: I'll Get By; A Ticket to Tomahawk
1951: Half Angel; Let's Make It Legal; Meet Me After the Show
1952: My Wife's Best Friend
1953: The Girl Next Door
1954: Fire Over Africa
1955: Gentlemen Marry Brunettes
1957: Abandon Ship!

SIDNEY SALKOW (1909-)

1937: Four Days' Wonder; Girl Overboard; Behind the Mike
1938: That's My Story; The Night Hawk; Storm over Bengal
1939: Fighting Thoroughbreds; Woman Doctor; Street of Missing Men; The Zero Hour; She Married a Cop; Flight at Midnight
1940: Cafe Hostess; The Lone Wolf Strikes; The Lone Wolf Meets a Lady; Girl from God's Country
1941: The Lone Wolf Keeps a Date; The Lone Wolf Takes a Chance; Time Out for Rhythm; Tillie the Toiler
1942: The Adventures of Martin Eden; Flight Lieutenant
1943: City Without Men; The Boy from Stalingrad
1946: Faithful in My Fashion
1947: Millie's Daughter; Bulldog Drummond at Bay
1948: Sound of the Avenger
1949: La Strada Buia (w/ Marino Girolami; Ital.)
1950: La Rivale dell'imperatrice (w/ Jacopo Comin; Ital.)
1952: Scarlet Angel; The Golden Hawk; The Pathfinder
1953: Prince of Pirates; Jack McCall, Desperado; Raiders of the Seven Seas
1954: Sitting Bull
1955: Robber's Roost; Las Vegas Shakedown; Shadow of the Eagle (Brit.); Toughest Man Alive
1956: Gun Brothers
1957: The Iron Sheriff; Gun Duel in Durango; Chicago Confidential

1960: The Big Night
1963: Twice Told Tales
1964: The Quick Gun; The Last Man on Earth (Ital.); Blood on the Arrow
1965: The Great Sioux Massacre
1966: The Murder Game

BILL SAMPSON (see JOHN ERMAN)

EDWARDS SAMPSON

1954: The Fast and the Furious

HENRI SAMUELS (see HARRY S. WEBB)

RAYMOND SAMUELS (see BERNARD B. RAY)

DENIS SANDERS (1929-)

1959: Crime and Punishment, U.S.A.
1962: War Hunt
1964: One Man's Way; Shock Treatment
1970: Elvis--That's the Way It Is (doc.)
1971: Soul to Soul (doc.; Ghana)
1973: Invasion of the Bee Girls

MARK SANDRICH (1900-1945)

1930: The Talk of Hollywood
1933: Melody Cruise; Aggie Appleby; Maker of Men
1934: Hips, Hips, Hooray; Cockeyed Cavaliers; The Gay Divorcee
1935: Top Hat
1936: Follow the Fleet; A Woman Rebels
1937: Shall We Dance
1938: Carefree
1939: Man About Town
1940: Buck Benny Rides Again; Love Thy Neighbor
1941: Skylark
1942: Holiday Inn
1943: So Proudly We Hail
1944: I Love a Soldier; Here Come the Waves

CLIFFORD SANFORTH

1938: I Demand Payment

ANTONIO SANTEAN

1964: The Glass Cage (ret. Den of Doom)

ALFRED SANTELL (1895-)

1930: The Arizona Kid; The Sea Wolf
1931: Body and Soul; Daddy Long Legs; Sob Sister
1932: Polly of the Circus; Rebecca of Sunnybrook Farm; Tess of the Storm Country
1933: Bondage; The Right to Romance
1934: The Life of Vergie Winters
1935: The Dictator (w/ Victor Saville; Brit.); People Will Talk; A Feather in Her Hat
1936: Winterset
1937: Interns Can't Take Money; Breakfast for Two
1938: Cocoanut Grove; Having a Wonderful Time; The Arkansas Traveler
1939: Our Leading Citizen
1941: Aloma of the South Seas
1942: Beyond the Blue Horizon
1943: Lack London
1944: The Hairy Ape
1945: Mexicana
1946: That Brennan Girl

CIRIO SANTIAGO

1973: Fly Me; Savage!
1975: TNT Jackson; Cover Girl Models
1976: Ebony Ivory and Jade; Hustler Squad; The Muthers

JOSEPH SANTLEY (1889-1971)

1929: The Cocoanuts (w/ R. Florey)
1930: Swing High
1934: The Loud Speaker; Young and Beautiful
1935: Million Dollar Baby; Harmony Lane; Waterfront Lady
1936: Dancing Feet; Her Master's Voice; Laughing Irish Eyes; The Harvester; We Went to College; Walking on Air; The Smartest Girl in Town
1937: Meet the Missus; There Goes the Groom
1938: She's Got Everything; Blonde Cheat; Always in Trouble; Swing, Sister, Swing
1939: Spirit of Culver; The Family Next Door; Two Bright Boys
1940: Music in My Heart; Melody and Moonlight; Melody Ranch; Behind the News

1941: Dancing on a Dime; Sis Hopkins; Rookies on Parade; Puddin'-head; Ice-Capades; Down Mexico Way
1942: A Tragedy at Midnight; Yokel Boy; Remember Pearl Harbor; Joan of Ozark; Call of the Canyon
1943: Chatterbox; Shantytown; Thumbs Up; Sleepy Lagoon; Here Comes Elmer
1944: Rosie, the Riveter; Jamboree; Goodnight Sweetheart; Three Little Sisters; Brazil
1945: Earl Carroll Vanities; Hitchhike to Happiness
1946: Shadow of a Woman
1949: Make Believe Ballroom
1950: When You're Smiling

RICHARD C. SARAFIAN (c.1927-)

1962: Terror at Black Falls
1965: Andy
1969: Run Wild, Run Free
1971: Vanishing Point; Fragment of Fear (Brit.); Man in the Wilderness
1973: Lolly Madonna XXX; The Man Who Loved Cat Dancing
1978: The Next Man
1979: Sunburn

JOSEPH SARGENT

1966: One Spy Too Many
1968: The Hell with Heroes
1970: Colossus--The Forbin Project
1972: The Man
1973: White Lightning
1974: The Taking of Pelham One Two Three
1977: MacArthur
1979: Goldengirl

MICHAEL SARNE (1939-) Brit.

(American films only)
1970: Myra Breckinridge

JOE SARNO

1968: Moonlighting Wives

PETER SAVAGE

1970: House in Naples

VICTOR SAVILLE (1897-1979) Brit.

(American films only)
1943: Forever and a Day (w/ others)
1945: Tonight and Every Night
1946: The Green Years
1947: Green Dolphin Street; If Winter Comes
1949: Conspirators (U.S./Brit.)
1950: Kim
1951: Calling Bulldog Drummond (U.S./ Brit.)
1954: The Long Wait
1955: The Silver Chalice

DAVID H. SAWYER

1970: Other Voices (doc.)

ARMAND SCHAEFER (1898-)

1931: Hurricane Horseman; Reckless Riders
1932: Law and the Lawless; Wyoming Whirlwind; Cheyenne Cyclone; Sinister Hands
1933: Terror Trail; Fighting Texans; Sagebrush Trail; Outlaw Justice
1934: Sixteen Fathoms Deep

(Serials)

1931: The Lightning Warrior (w/ Ben Kline)
1932: The Hurricane Express (w/ J. P. McGowan)
1933: Fighting with Kit Carson (w/ Colbert Clark); The Three Musketeers (w/ C. Clark)
1934: Burn 'Em Up Barnes (w/ C. Clark); The Law of the Wild (w/ B. Eason); The Lost Jungle (w/ David Howard)
1935: The Miracle Rider (w/ B. Eason)

GEORGE SCHAEFER (1920-)

1963: Macbeth
1969: Pendulum; Generation
1971: Doctors' Wives
1973: Once Upon a Scoundrel
1979: An Enemy of the People

FRANKLIN SCHAFFNER (1920-)

1963: The Stripper
1964: The Best Man

1965: The War Lord
1968: Planet of the Apes; The Double Man
1970: Patton
1971: Nicholas and Alexandra (Brit.)
1973: Papillon
1977: Islands in the Stream
1978: The Boys from Brazil

DON SCHAIN

1972: A Place Called Today; The Abductors
1973: Girls Are for Loving

DORE SCHARY (1905-)

1963: Act One

JERRY SCHATZBERG

1970: Puzzle of a Downfall Child
1971: The Panic in Needle Park
1973: Scarecrow
1976: Sweet Revenge
1979: The Seduction of Joe Tynan

ROBERT SCHEERER

1970: Adam at 6 A.M.
1973: The World's Greatest Athlete

HENNING SCHELLERUP

1973: Sweet Jesus, Preacher Man
1979: In Search of Historic Jesus; Beyond Death's Door; Legend of Sleepy Hollow

GEORGE SCHENCK

1972: Superbeast

HARRY SCHENCK

1934: Beyond Bengal (doc.)

VICTOR SCHERTZINGER (1880-1941)

1929: Redskin; Nothing but the Truth; The Wheel of Life; Fashions in Love
1930: The Laughing Lady; Paramount on Parade (w/ others); Safety in Numbers; Heads Up
1931: The Woman Between; Friends and Lovers
1932: Strange Justice; Uptown New York
1933: The Constant Woman; The Cocktail Hour; My Woman
1934: Beloved; One Night of Love
1935: Let's Live Tonight; Love Me Forever
1936: The Music Goes 'Round
1937: Something to Sing About
1939: The Mikado (Brit.)
1940: Road to Singapore; Rhythm on the River
1941: Road to Zanzibar; Kiss the Boys Goodbye; Birth of the Blues
1942: The Fleet's In

TOM SCHEUER

1974: Gosh

LAWRENCE SCHILLER

1971: The Lexington Experience (doc.); The American Dreamer

GEORGE SCHLATTER

1976: Norman ... Is That You?

JOHN SCHLESINGER (1926-) Brit.

1962: A King of Loving (Brit.)
1963: Billy Liar (Brit.)
1965: Darling (Brit.)
1967: Far from the Madding Crowd (Brit.)
1969: Midnight Cowboy
1971: Sunday, Bloody Sunday (Brit.)
1973: Visions of Eight (doc. w/ others)
1975: The Day of the Locust
1976: Marathon Man
1979: Yanks

THOMAS SCHMIDT

1973: Hot Summer Week

DAVID SCHMOELLER

1979: The Tourist Trap

ROBERT ALLEN SCHNITZER

1975: No Place to Hide
1976: The Premonition

ERNEST B. SCHOEDSACK (1893-)

1929: Four Feathers (w/ Merian C. Cooper, Lother Mendes)
1931: Rango
1932: The Most Dangerous Game (w/ Irving Pichel)
1933: King Kong (w/ Merian C. Cooper); Blind Adventure; Son of Kong
1934: Long Lost Father
1935: The Last Days of Pompeii
1937: Trouble in Morocco; Outlaws of the Orient
1940: Dr. Cyclops
1949: Mighty Joe Young

JACK SCHOLL

1950: Holiday Rhythm

WILLIAM SCHORR

1932: Forgotten Commandments (w/ L. Gasnier)

WAYNE A. SCHOTTEN

1970: Friday on My Mind

PAUL SCHRADER (1946-)

1978: Blue Collar
1979: Hardcore; American Gigolo

MICHAEL SCHULTZ (1938-)

1973: Together for Days
1974: Honeybaby, Honeybaby
1975: Cooley High
1976: Car Wash
1977: Which Way Is Up?; Greased Lightning

1978: Sgt. Pepper's Lonely Hearts Club Band
1979: Scavenger Hunt

REINHOLD SCHUNZEL (1886-1954) Germ.

(American films only)
1938: Rich Man, Poor Girl
1939: Balalaika; Ice Follies of 1939
1941: New Wine

HAROLD D. SCHUSTER (1902-   )

1937: Dinner at the Ritz; Wings of the Morning (Brit.)
1938: Swing That Cheer; Exposed
1939: One Hour to Live
1940: Zanzibar; Ma! He's Making Eyes at Me; South to Karanga; Diamond Frontier; Framed
1941: A Very Young Lady; Small Town Deb; A Modern Monte Cristo
1942: Girl Trouble; The Postman Didn't Ring; On the Sunny Side
1943: My Friend Flicka
1944: Marine Raiders
1946: Breakfast in Hollywood
1947: The Tender Years
1948: So Dear to My Heart
1952: Kid Monk Baroni
1953: Jack Slade
1954: Loophole; Port of Hell; Security Risk
1955: Finger Man; The Return of Jack Slade; Tarzan's Hidden Jungle
1957: Courage of Black Beauty; Dragoon Wells Massacre; Portland Expose

LAURENCE SCHWAB

1930: Follow Thru (w/ Lloyd Corrigan)
1933: Take a Chance (w/ Monte Brice)

DOUGLAS SCHWARTZ

1971: The Peace Killers
1973: Your Three Minutes Are Up

MARTIN SCORSESE

1968: Who's That Knocking at My Door?
1970: Street Scenes, 1970 (doc.)
1972: Boxcar Bertha

1973: Mean Streets
1975: Alice Doesn't Live Here Any More
1976: Taxi Driver
1977: New York, New York
1978: The Last Waltz

EWING SCOTT

1932: Igloo (re-ed. as Red Snow & credited as 2nd Unit Director)
1934: Renegade
1937: Hollywood Cowboy; Windjammer; Headin' West; Hollywood Round-Up
1938: Man Hunters of the Caribbean (doc. w/ A. Roosevelt)
1947: Untamed Fury
1948: Harpoon
1949: Arctic Manhunt

GEORGE C. SCOTT (1927-)

1972: Rage
1974: The Savage Is Loose

RIDLEY SCOTT

1978: The Duellists (Brit.)
1979: Alien

SHERMAN SCOTT (see SAM NEWFIELD)

AUBREY H. SCOTTO

1932: Divorce Racket; Uncle Moses
1934: Three Loves; I Hate Women
1935: $1,000 a Minute; Hitch Hike Lady; Smart Girl
1936: Ticket to Paradise; Follow Your Heart; Happy Go Lucky; Palm Springs
1937: Blazing Barriers
1938: Gambling Ship; I Was a Criminal; Little Miss Roughneck

FRED F. SEARS (1913-1957)

1949: Desert Vigilante; Horsemen of the Sierras
1950: Across the Badlands; Raiders of Tomahawk Creek; Lightning Guns
1951: Prairie Roundup; Ridin' the Outlaw Trail; Snake River Desperadoes; Bonanza Town; Pecos River

1952: Smokey Canyon; The Hawk of Wild River; The Kid from Broken Gun; Last Train from Bombay; Target Hong Kong
1953: Ambush at Tomahawk Gap; The 49th Man; Mission over Korea; Sky Commando; The Nebraskan; El Alamein
1954: Overland Pacific; The Miami Story; Massacre Canyon; The Outlaw Stallion
1955: Wyoming Renegades; Cell 2455, Death Row; Chicago Syndicate; Apache Ambush; Teen-age Crime Wave; Inside Detroit
1956: Fury at Gunsight Pass; Rock Aroung the Clock; Earth vs. the Flying Saucers; The Werewolf; Miami Expose; Cha-Cha-Cha Boom!; Rumble on the Docks; Don't Knock the Rock
1957: Utah Blaine; Calypso Heat Wave; The Night the World Exploded; The Giant Claw; Escape from San Quentin
1958: The World Was His Jury; Going Steady; Crash Landing; Badman's Country; Ghost of the China Sea

VICTOR SEASTROM (1879-1960) Swed.

1930: A Lady to Love; Father and Son (Swed.)
1937: Under the Red Robe (Brit.)

GEORGE SEATON (1911-   )

1945: Diamond Horseshoe; Junior Miss
1946: The Shocking Miss Pilgrim
1947: Miracle on 34th Street
1948: Apartment for Peggy
1949: Chicken Every Sunday
1950: The Big Lift; For Heaven's Sake
1952: Anything Can Happen
1953: Little Boy Lost
1954: The Country Girl
1956: The Proud and the Profane
1957: Williamsburg: The Story of a Patriot (doc.)
1958: Teacher's Pet
1961: The Pleasure of His Company
1962: The Counterfeit Traitor
1963: The Hook
1965: 36 Hours
1968: What's So Bad About Feeling Good?
1970: Airport
1973: Showdown

BEVERLY & FERD SEBASTIAN

1974: Gator Bait; The Single Girls
1979: Delta Fox; On the Air Live with Captain Midnight

EDWARD SEDGWICK (1893-1953)

1928: Circus Rookies
1929: Spite Marriage
1930: Free and Easy; Dough Boys
1931: Parlor, Bedroom and Bath; A Dangerous Affair; Maker of Men
1932: The Passionate Plumber; Speak Easily
1933: What! No Beer?; Saturday's Millions; Horseplay
1934: The Poor Rich; I'll Tell the World; Here Comes the Groom; Death on the Diamond
1935: Father Brown, Detective; Murder in the Fleet; The Virginia Judge
1936: Mister Cinderella
1937: Pick a Star; Riding on Air; Fit for a King
1938: The Gladiator
1939: Burn 'Em up O'Connor; Beware, Spooks!
1940: So You Won't Talk?
1943: Air Raid Wardens
1948: A Southern Yankee
1951: Ma and Pa Kettle Back on the Farm

GERRI SEDLEY

1975: Teenage Hitchhikers

ALEX SEGAL (1915-1977)

1956: Ransom
1963: All the Way Home
1965: Joy in the Morning; Harlow

LEWIS R. SEILER (1891-1964)

1929: The Ghost Talks; Girls Gone Wild; A Song of Kentucky
1932: No Greater Love
1933: Deception
1934: Frontier Marshal
1935: Charlie Chan in Paris; Asegure a su Mujer (Span.); Ginger; Paddy O'Day
1936: Here Comes Trouble; The First Baby; Star for a Night; Career Woman
1937: Turn Off the Moon
1938: He Couldn't Say No; Crime School; Penrod's Double Trouble; Heart of the North
1939: King of the Underworld; You Can't Get Away with Murder; The Kid from Kokomo; Hell's Kitchen (w/ E. A. Dupont); Dust Be My Destiny
1940: It All Came True; Flight Angels; Murder in the Air; Tugboat Annie Sails Again; South of Suez

1941: Kisses for Breakfast; The Smiling Ghost; You're in the Army Now
1942: The Big Shot; Pittsburgh
1943: Guadalcanal Diary
1944: Something for the Boys
1945: Molly and Me; Doll Face
1946: If I'm Lucky
1948: Whiplash
1950: Breakthrough
1951: The Tanks Are Coming
1952: The Winning Team; Operation Secret
1953: The System
1954: The Bamboo Prison
1955: Women's Prison
1956: Battle Stations; Over-Exposed
1958: The True Story of Lynn Stuart

HEINZ SEILMANN

1974: Vanishing Wilderness (w/ Arthur Dubs)

WILLIAM A. SEITER (1891-1964)

1929: Synthetic Sin; Why Be Good?; Smiling Irish Eyes; Prisoners; Footlights and Fools
1930: Strictly Modern; Back Pay; The Flirting Widow; The Love Racket; The Truth About Youth; Sunny
1931: Kiss Me Again; Going Wild; Big Business Girl; Too Many Cooks; Caught Plastered; Peach O'Reno; Full of Notions
1932: Way Back Home; Girl Crazy; Young Bride; Is My Face Red?; Hot Saturday; If I Had a Million (w/ others)
1933: Hello, Everybody!; Diplomaniacs; Professional Sweetheart; Chance at Heaven
1934: Sons of the Desert; Rafter Romance; Sing and Like It; Love Birds; We're Rich Again; The Richest Girl in the World
1935: Roberta; The Daring Young Man; Orchids to You; In Person; If You Could Only Cook
1936: The Moon's Our Home; The Case Against Mrs. Ames; Dimples; Stowaway
1937: This Is My Affair; The Life of the Party; Life Begins in College
1938: Sally, Irene and Mary; Three Blind Mice; Room Service; Thanks for Everything
1939: Susannah of the Mounties; Allegheny Uprising
1940: It's a Date; Hired Wife
1941: Nice Girl?; Appointment for Love
1942: Broadway; You Were Never Lovelier
1943: Four Jills in a Jeep; Destroyer; A Lady Takes a Chance
1944: Belle of the Yukon
1945: It's a Pleasure; The Affairs of Susan; That Night with You
1946: Little Giant; Lover Come Back

1947: I'll Be Yours
1948: Up in Central Park; One Touch of Venus
1950: Borderline
1951: Dear Brat
1953: The Lady Wants Mink; Champ for a Day
1954: Make Haste to Live

GEORGE B. SEITZ (1888-1944)

1929: Black Magic
1930: The Murder on the Roof; Guilty; Midnight Mystery; Danger Lights
1931: Drums of Jeopardy; The Lion and the Lamb; Arizona; Shanghaied Love; The Night Beat
1932: Docks of San Francisco; Sally of the Subway; Sin's Pay Day; Passport to Paradise; The Widow in Scarlet
1933: Treason; The Thrill Hunter; The Women in His Life
1934: Lazy River; The Fighting Rangers
1935: Only Eight Hours; Society Doctor; Shadow of Doubt; Times Square Lady; Calm Yourself; Woman Wanted; Kind Lady
1936: Exclusive Story; Absolute Quiet; The Three Wise Guys; The Last of the Mohicans (w/ Wallace Fox); Mad Holiday
1937: Under Cover of Night; A Family Affair; The Thirteenth Chair; Mama Steps Out; Between Two Women; My Dear Miss Aldrich
1938: You're Only Young Once; Judge Hardy's Children; Yellow Jack; Love Finds Andy Hardy; Out West with the Hardys
1939: The Hardys Ride High; Six Thousand Enemies; Thunder Afloat; Judge Hardy and Son
1940: Andy Hardy Meets Debutante; Kit Carson; Sky Murder; Gallant Sons
1941: Andy Hardy's Private Secretary; Life Begins for Andy Hardy
1942: A Yank on the Burma Road; The Courtship of Andy Hardy; Pierre of the Plains; Andy Hardy's Double Life
1944: Andy Hardy's Blonde Trouble

STEVE SEKELY (ISTVAN SZEKELY) (1899-1979) Hung.

(American films only)
1940: A Miracle on Main Street
1943: Behind Prison Walls; Revenge of the Zombies; Women in Bondage
1944: Lady in the Death House; Waterfront; My Buddy; Lake Placid Serenade
1946: The Fabulous Suzanne
1947: Blonde Savage
1948: Hollow Triumph (alt. The Scar)
1949: Amazon Quest
1952: Stronghold
1957: Cartouche (Ital.)
1959: Desert Desperadoes (Ital.)

1963: The Day of the Triffids (Brit.)
1969: Kenner

LESLEY SELANDER (1900-1980)

1936: Sandflow; Ride 'Em Cowboy; Empty Saddles; Boss Rider of Gun Creek
1937: Left-Handed Law; Smoke Tree Range; Hopalong Rides Again; The Barrier; Partners of the Plains
1938: Cassidy of Bar 20; The Heart of Arizona; Bar 20 Justice; Pride of the West; The Mysterious Rider; Sunset Trail; The Frontiersmen
1939: Silver on the Sage; Heritage of the Desert; Renegade Trail; Range War
1940: Santa Fe Marshal; Knights of the Range; The Light of Western Stars; Hidden Gold; Stagecoach War; Three Men from Texas; Cherokee Strip
1941: Doomed Caravan; The Round Up; Pirates on Horseback; Wide Open Town; Riders of the Timberline; Stick to Your Guns; Thundering Hoofs
1942: Undercover Man; Bandit Ranger; Red River Robin Hood
1943: Lost Canyon; Buckskin Frontier; Border Patrol; Colt Comrades; Bar 20; Riders of the Deadline
1944: Lumberjack; Call of the Rockies; Forty Thieves; Bordertown Trail; Sheriff of Sundown; Firebrands of Arizona; Sheriff of Las Vegas; Stagecoach to Monterey; Cheyenne Wildcat
1945: The Great Stagecoach Robbery; The Vampire's Ghost; The Trail of Kit Carson; Three's a Crowd; The Fatal Witness; Phantom of the Plains
1946: The Catman of Paris; Passkey to Danger; Traffic in Crime; Night Train to Memphis; Out California Way
1947: The Pilgrim Lady; The Last Frontier Uprising; Saddle Pals; Robin Hood of Texas; The Red Stallion; Blackmail
1948: Panhandle; Guns of Hate; Belle Starr's Daughter; Indian Agent; Strike It Rich
1949: Brothers in the Saddle; Rustlers; Stampede; Sky Dragon; The Mysterious Desperado; Masked Raiders; Riders of the Range
1950: Dakota Lil; Storm over Wyoming; Rider from Tucson; The Kangaroo Kid (Australia); Rio Grande Patrol; Short Grass; Law of the Badlands
1951: Saddle Legion; I Was an American Spy; Gunplay; Cavalry Scout; Pistol Harvest; The Highwayman; Flight to Mars; Overland Telegraph
1952: Fort Osage; Trail Guide; Road Agent; Desert Passage; The Raiders; Battle Zone; Flat Top
1953: Fort Vengeance; Cow Country; War Paint; Fort Algiers; The Royal African Rifles; Fighter Attack
1954: Shotgun; Tall Man Riding; Desert Sands; Fort Yuma
1956: Tomahawk Trail; Revolt at Fort Laramie; Outlaw's Son; Taming Sutton's Gal; The Wayward Girl
1958: The Lone Ranger and the Lost City of Gold

1965: War Party; Fort Courageous; Convict Stage; Town Tamer
1966: The Texican
1967: Fort Utah
1968: Arizona Bushwackers

(Serials)

1945: Jungle Raiders

DAVID SELMAN

1934: The Prescott Kid; The Westerner
1935: Fighting Shadows; Gallant Defender; Justice on the Range; The Revenge Rider; Riding Wild; Square Shooter
1936: Shakedown; Killer at Large; The Cowboy Star; The Mysterious Avenger; Secret Patrol; Dangerous Intrigue; Tugboat Princess
1937: The Texas Trail; Find the Witness
1938: Woman Against the World

EDGAR SELWYN (1875-1944)

1930: War Nurse; Girl in the Show
1931: Men Call It Love; The Sin of Madelon Claudet
1932: Skyscraper Souls
1933: Men Must Fight; Turn Back the Clock
1934: The Mystery of Mr. X

MACK SENNETT (1880-1960)

1929: Midnight Daddies
1933: Hypnotized

MARIO SEQUI

1968: The Cobra

JOHN SEVERSON

1971: Pacific Vibrations (doc.)

NICHOLAS SGARRO

1975: The Happy Hooker

JOSEF SHAFTEL

1956: The Naked Hills; No Place to Hide

KRISHNA SHAH

1972: Rivals
1976: The River Niger

BERNARD SHAKEY

1979: Rust Never Sleeps (doc.)

FRANKLIN SHAMROY (see BERNARD B. RAY)

MAXWELL SHANE (1905-)

1947: Fear in the Night
1949: City Across the River
1953: The Glass Wall
1955: The Naked Street
1956: Nightmare

KEN SHAPIRO

1974: The Groove Tube

MEL SHAPIRO

1979: Sammy Stops the World

STEFAN SHARFF

1965: Across the River

ROBERT K. SHARPE

1970: Before the Mountain Was Moved (doc.)

MELVILLE SHAVELSON (1917-)

1955: The Seven Little Foys
1957: Beau James

1958: Houseboat
1959: The Five Pennies
1960: It Started in Naples (Ital.)
1961: On the Double
1962: The Pigeon That Took Rome (Ital.)
1963: A New Kind of Love (Fr.)
1966: Cast a Giant Shadow
1968: Yours, Mine and Ours
1972: The War Between Men and Women
1974: Mixed Company

JACK SHEA

1968: Dayton's Devils
1969: The Monitors

WILLIAM SHEA

1936: Girl of the Ozarks

BARRY SHEAR (1920?-1979)

1968: Wild in the Streets
1971: The Todd Killings
1972: Across 110th Street
1973: The Deadly Trackers (Span.)

DONALD SHEBIB

1970: Goin' Down the Road
1972: Rip-Off

FORREST SHELDON

1931: Law of the Rio Grande (w/ Bennett Cohen)
1932: Between Fighting Men; Dynamite Ranch; Hellfire Austin; Lone Trail (from serial The Sign of the Wolf)
1935: Wilderness Mail

(Serial)

1931: The Sign of the Wolf (w/ Harry Webb)

NORMAN SHELDON

1949: Rio Grande

SIDNEY SHELDON (1917-)

1953: Dream Wife
1957: The Buster Keaton Story

JOSHUA SHELLEY

1967: The Perils of Pauline (w/ H. Leonard)

WALTER SHENSON (1919-)

1971: Welcome to the Club

ANTONIO SHEPHERD

1979: Chorus Call

JACK SHER (1913-)

1956: Four Girls in Town
1958: Kathy O'
1959: The Wild and the Innocent
1960: The Three Worlds of Gulliver (Brit.)
1961: Love in a Goldfish Bowl

JAY J. SHERIDAN

1966: Nashville Rebel

EDWIN SHERIN

1971: Valdez Is Coming
1972: My Old Man's Place

GEORGE SHERMAN (1908-)

1938: The Purple Vigilantes; Wild Horse Rodeo; Outlaws of Sonora; Riders of the Black Hills; Heroes of the Hills; Pals of the Saddle; Overland Stage Raiders; Rhythm of the Saddle; Santa Fe Stampede; Red River Range
1939: Mexicali Rose; The Night Riders; Three Texas Steers; Wyoming Outlaw; Colorado Sunset; New Frontier; The Kansas Terrors; Rovin' Tumbleweeds; Cowboys from Texas; South of the Border
1940: Ghost Valley Raiders; Covered Wagon Days; Rocky Mountain

Rangers; One Man's Law; The Tulsa Kid; Under Texas Skies; The Trail Blazers; Texas Terrors; Lone Star Raiders
1941: Wyoming Wildcat; The Phantom Cowboy; Two-Gun Sheriff; Desert Bandit; Kansas Cyclone; Citadel of Crime; The Apache Kid; Death Valley Outlaws; A Missouri Outlaw
1942: Arizona Terrors; Stagecoach Express; Jesse James, Jr.; The Cyclone Kid; The Sombrero Kid; X Marks the Spot; London Blackout Murders
1943: The Purple V; The Mantrap; False Faces; The West Side Kid; A Scream in the Dark; Mystery Broadcast
1944: The Lady and the Monster; Storm over Lisbon
1945: The Crime Doctor's Courage
1946: The Bandit of Sherwood Forest (w/ Henry Levin); Talk About a Lady; Renegades; The Gentleman Misbehaves; Personality Kid; Secret of the Whistler
1947: Last of the Redmen
1948: Relentless; Black Bart; River Lady; Feudin', Fussin' and A-Fightin'; Larceny
1949: Red Canyon; Calamity Jane and Sam Bass; Yes Sir, That's My Baby; Sword in the Desert
1950: Comanche Territory; Spy Hunt; The Sleeping City
1951: Tomahawk; Target Unknown; The Golden Horde; The Raging Tide
1952: Steel Town; The Battle at Apache Pass; Back at the Front; Against All Flags
1953: The Lone Hand; The Veils of Bagdad; War Arrow
1954: Border River; Johnny Dark; Dawn at Socorro
1955: Chief Crazy Horse; The Treasure of Pancho Villa; Count Three and Pray
1956: Comanche; Reprisal!
1957: The Hard Man
1958: The Last of the Fast Guns; Ten Days to Tulara
1959: The Son of Robin Hood (Brit.); The Flying Fontaines
1960: Hell Bent for Leather; For the Love of Mike; The Enemy General; The Wizard of Baghdad
1961: The Fiercest Heart
1964: Panic Button (Ital.)
1965: Murieta (Span.)
1966: Smoky
1971: Big Jake

LOWELL SHERMAN (1885-1934)

1930: The Pay-Off; Lawful Larceny
1931: Bachelor Apartment; The Royal Bed; High Stakes
1932: The Greeks Had a Word for Them; False Faces; Ladies of the Jury
1933: Morning Glory; She Done Him Wrong; Broadway Thru a Keyhole
1934: Born to Be Bad
1935: Night Life of the Gods

VINCENT SHERMAN (1906-)

1939: The Return of Dr. X
1940: Saturday's Children; The Man Who Talked Too Much
1941: Flight From Destiny; Underground
1942: All Through the Night; The Hard Way
1943: Old Acquaintance
1944: In Our Time; Mr. Skeffington
1945: Pillow to Post
1946: Janie Gets Married
1947: Nora Prentiss; The Unfaithful
1948: The Adventures of Don Juan
1949: The Hasty Heart (Brit.)
1950: The Damned Don't Cry; Backfire; Harriet Craig
1951: Goodbye, My Fancy
1952: Lone Star; Affair in Trinidad
1957: The Garment Jungle
1958: Naked Earth (Africa)
1959: The Young Philadelphians
1960: Ice Palace
1961: A Fever in the Blood; The Second Time Around
1968: Cervantes (Span.)

JOHN SHERWOOD

1956: The Creature Walks Among Us; Raw Edge
1957: The Monolith Monsters

PAT SHIELDS

1973: Frasier, the Sensuous Lion

LEE SHOLEM

1949: Tarzan's Magic Fountain
1950: Tarzan and the Slave Girl
1952: The Redhead from Wyoming
1953: The Stand at Apache River
1954: Cannibal Attack; Jungle Man-Eaters; Tobor the Great
1955: Ma and Pa Kettle at Waikiki
1956: Crime Against Joe; Emergency Hospital
1957: Hell Ship Mutiny (w/ Elmo Williams); The Pharaoh's Curse; Sierra Stranger

SIG SHORE

1975: Shining Star; That's the Way of the World

LYNN SHORES

1929: The Delightful Rogue
1936: The Glory Trail; Million to One; Rebellion; Women in Distress
1937: The Shadow Strikes; Here's Flash Casey
1940: Charlie Chan at the Wax Museum
1941: Golden Hoofs

SHERRY SHOURDS

1948: The Big Punch

HERMAN SHUMLIN (1898-1979)

1943: Watch on the Rhine
1945: Confidential Agent

MELVILLE SHYER

1933: Sucker Money (w/ Dorothy Reid); Murder in the Museum
1934: The Road to Ruin (w/ Dorothy Reid)

ANDY SIDARIS

1973: Stacey!
1979: Seven

GEORGE SIDNEY (1916-)

1941: Free and Easy
1942: Pacific Rendezvous
1943: Pilot No. 5; Thousands Cheer
1944: Bathing Beauty
1945: Anchors Aweigh
1946: The Harvey Girls; Holiday in Mexico
1947: Cass Timberlane
1948: The Three Musketeers
1949: The Red Danube
1950: Annie Get Your Gun; Key to the City
1951: Show Boat
1952: Scaramouche
1953: Kiss Me Kate; Young Bess
1955: Jupiter's Darling
1956: The Eddie Duchin Story
1957: Jeanne Eagels; Pal Joey
1960: Pepe; Who Was That Lady?
1963: A Ticklish Affair; Bye, Bye Birdie

1964: Viva Las Vegas
1966: The Swinger
1968: Half a Sixpence (Brit.)

DON SIEGEL (1912-)

1946: The Verdict
1949: The Big Steal; Night unto Night
1952: The Duel at Silver Creek; No Time for Flowers
1953: China Venture; Count the Hours
1954: Private Hell 36; Riot in Cell Block 11
1955: An Annapolis Story
1956: Crime in the Streets; Invasion of the Body Snatchers
1957: Baby Face Nelson
1958: The Gun Runners; The Line-Up; Spanish Affair (Span.)
1959: Hound Dog Man
1960: Edge of Eternity; Flaming Star
1962: Hell Is for Heroes
1964: The Killers
1967: Strangers on the Run
1968: Coogan's Bluff; Madigan
1969: Death of a Gunfighter (w/ Robert Totten)
1970: Two Mules for Sister Sara
1971: The Beguiled; Dirty Harry
1973: Charley Varrick
1974: The Black Windmill (Brit.)
1976: The Shootist
1977: Telefon
1979: Escape from Alcatraz

GLENN SILBER

1979: The War at Home (doc.)

ALF SILLIMAN, JR.

1970: The Stewardesses

JOAN MICKLIN SILVER

1975: Hester Street
1977: Between the Lines
1979: Head Over Heels

MARCEL SILVER

1929: William Fox Movietone Follies of 1929 (w/ David Butler); Married in Hollywood
1930: One Mad Kiss

RAPHAEL D. SILVER

1979: On the Yard

ELLIOT SILVERSTEIN (1927-)

1962: Belle Sommers
1965: Cat Ballou
1967: The Happening
1969: A Man Called Horse
1974: Deadly Honeymoon
1977: The Car

FRANCIS SIMON

1978: The Chicken Chronicles

FRANK SIMON

1968: The Queen (doc.)

S. SYLVAN SIMON (1910-1951)

1937: A Girl with Ideas; A Prescription for Romance
1938: The Crime of Dr. Hallet; Nurse from Brooklyn; The Road to Reno; Spring Madness
1939: Four Girls in White; The Kid from Texas; These Glamour Girls; Dancing Co-Ed
1940: Two Girls on Broadway; Sporting Blood; Dulcy
1941: Keeping Company; Washington Melodrama; Whistling in the Dark; The Bugle Sounds
1942: Rio Rita; Grand Central Murder; Tish; Whistling in Dixie
1943: Salute to the Marines; Whistling in Brooklyn
1944: Son of Lassie; Abbott and Costello in Hollywood
1946: Bad Bascomb; The Cockeyed Miracle; The Thrill of Brazil
1947: Her Husband's Affairs
1948: I Love Trouble; The Fuller Brush Man
1949: Lust for Gold

FRANK SINATRA (1915-)

1965: None but the Brave

ROBERT B. SINCLAIR (1905-1970)

1938: Dramatic School; Woman Against Woman
1939: Joe and Ethel Turp Call on the President

1940: The Captain Is a Lady; And One Was Beautiful
1941: The Wild Man of Borneo; I'll Wait for You; Down in San Diego
1942: Mr. and Mrs. North
1947: Mr. District Attorney
1948: That Wonderful Urge

ALEXANDER SINGER (1932-   )

1961: A Cold Wind in August
1964: Psyche 59 (Brit.)
1965: Love Has Many Faces
1971: Captain Apache
1972: Glass Houses

CURT SIODMAK (1902-   ) Germ.

(American films only)
1951: Bride of the Gorilla
1952: The Magnetic Monster
1957: Love Slaves of the Amazon
1959: Ski Fever (Czech./U.S./W. Germ./Pol./Austrian)

ROBERT SIODMAK (1900-1973) Germ.

(American films only)
1941: West Point Widow
1942: Fly by Night; The Night Before the Divorce; My Heart Belongs to Daddy
1943: Someone to Remember; Son of Dracula
1944: Phantom Lady; Cobra Woman; Christmas Holiday; The Suspect
1945: Uncle Harry; The Spiral Staircase
1946: The Killers; The Dark Mirror
1947: Time Out of Mind
1948: Cry of the City
1949: Criss Cross; The Great Sinner; Thelma Jordan; The Crimson Pirate
1950: Deported
1951: The Whistle at Eaton Falls
1968: Custer of the West (Span/U.S.)

DOUGLAS SIRK (1900-   ) Danish

(American films only)
1935: All That Heaven Allows
1943: Hitler's Madman
1944: Summer Storm
1946: Thieves' Holiday; Scandal in Paris

1947: Lured
1948: Sleep My Love
1949: Shockproof; Slightly French
1950: Mystery Submarine
1951: The First Legion; The Lady Pays Off; Thunder on the Hill; Weekend with Father
1952: Has Anybody Seen My Gal?; Meet Me at the Fair; No Room for the Groom
1953: All I Desire; Take Me to Town
1954: Magnificent Obsession; Sign of the Pagan; Taza, Son of Cochise
1955: Captain Lightfoot; All That Heaven Allows
1956: Battle Hymn; There's Always Tomorrow; Written on the Wind
1957: Interlude; The Tarnished Angels (ret. Pylon)
1958: A Time to Love and a Time to Die
1959: Imitation of Life

LANE SLATE

1971: Clay Pigeon (w/ Tom Stern)

ROBERT F. SLATZER

1973: Big Foot

JOHN SLEDGE

1958: New Orleans After Dark; Invisible Avenger (w/ J. W. Howe)

PAUL H. SLOANE (1893-)

1929: Hearts in Dixie (w/ A. H. Van Buren)
1930: The Cuckoos; Half Shot at Sunrise; Three Sisters
1931: Traveling Husbands; Consolation Marriage
1932: War Correspondent
1933: The Woman Accused; Terror Abroad
1934: Straight Is the Way; The Lone Cowboy; Down to Their Last Yacht
1935: Here Comes the Band
1940: Geronimo!
1951: The Sun Sets at Dawn

EDWARD SLOMAN (1887-)

1929: The Girl on the Barge; The Kibitzer
1930: The Lost Zeppelin; Puttin' on the Ritz; Hell's Island; Soldiers and Women
1931: His Woman; The Conquering Horde; Gun Smoke; Murder by the Clock; Caught

1932: Wayward
1934: There's Always Tomorrow
1935: A Dog of Flanders
1938: The Jury's Secret

JACK SMIGHT (1926-)

1964: I'd Rather Be Rich
1965: The Third Day
1966: Harper; Kaleidoscope (Brit.)
1968: No Way to Treat a Lady; The Secret War of Harry Frigg
1969: The Illustrated Man; Strategy of Terror
1970: Rabbit, Run; The Traveling Executioner
1974: Airport 1975
1976: Midway
1977: Damnation Alley
1979: Fast Break

CLIFFORD SMITH (1894-1937)

1932: Riders of Golden Gulch; The Texan
1935: Devil's Canyon; Five Bad Men

(Serials)

1936: Ace Drummond (w/ F. Beebe); The Adventures of Frank Merriwell
1937: Jungle Jim; Radio Patrol; Secret Agent X-9; Wild West Days (all w/ F. Beebe)

HARRY W. SMITH

1953: Louisiana Territory

HOWARD SMITH

1972: Marjoe (doc. w/ Sarah Kernochan)
1977: Gizmo (doc.)

NOEL MASON SMITH (1890?-)

1929: Bachelor's Club; Back from Shanghai
1930: Heroic Lover
1931: Yankee Don; Dancing Dynamite; Scareheads
1935: Fighting Pilot
1936: Trailin' West; King of Hockey
1937: California Mail; Guns of the Pecos; Blazing Sixes; The Cherokee Strip; Over the Goal
1938: Mystery House

1939: Secret Service of the Air; Code of the Secret Service; Torchy Plays with Dynamite; Cowboy Quarterback
1940: Ladies Must Live; Calling All Husbands; Always a Bride; Father Is a Prince
1941: The Case of the Black Parrot; The Nurse's Secret; Burma Convoy
1952: Cattle Town

(Serial)

1942: Gang Busters (w/ Ray Taylor)

PAUL GERARD SMITH

1940: Margie (w/ O. Garrett); Sandy Gets Her Man (w/ O. Garrett)

ALLEN SMITHEE (pseud. for DON SIEGEL)

ROBERT R. SNODY

1930: Love Kiss

ROBERT SNYDER

1953: Blood Brothers (doc.)
1974: Anaïs Nin Observed (doc.); World of Buckminster Fuller (doc.); The Henry Miller Odyssey (doc.)

HENRY SOKAL

1937: A Smile in the Storm (w/ B. Vorhaus)
1940: They Met on Skis

PAUL SOLDAY

1973: The Devil's Wedding Night

PETER SOLMO

1976: Jack and the Beanstalk (anim.)

GEORGE SOMNES

1933: Girl in 419 (w/ Alexander Hall); Midnight Club (w/ A. Hall); Torch Singer (w/ A. Hall)
1934: Wharf Angel (w/ W. C. Menzies)

LARRY SPANGLER

1973: The Soul of Nigger Charley
1974: A Knife for the Ladies
1976: Joshua

ROBERT SPARR

1969: More Dead Than Alive; Once You Kiss a Stranger

NORMAN SPENCER

1935: Rainbow's End

KAREN SPERLING

1971: Make a Face

STEVEN SPIELBERG (1947-)

1974: The Sugarland Express
1975: Jaws
1977: Close Encounters of the Third Kind
1979: 1941

WILLIAM SPIER

1952: Lady Possessed (w/ R. Kellino)

MARTIN J. SPINELLI

1973: Childhood II (doc.)

NAT SPITZER

1931: Monsters of the Deep

G. D. SPRADLIN

1972: The Only Way Home

CHANDLER SPRAGUE

1930: Not Damaged; The Dancers
1931: Their Mad Moment (w/ H. MacFadden)

R. G. SPRINGSTEEN (1904-   )

1945: Marshal of Laredo; Colorado Pioneers; Wagon Wheels Westward
1946: California Gold Rush; Sheriff of Redwood Valley; Home on the Range; Sun Valley Cyclone; Man from Rainbow Valley; Conquest of Cheyenne; Santa Fe Uprising; Stagecoach to Denver
1947: Vigilantes of Boomtown; Homesteaders of Paradise Valley; Oregon Trail Scouts; Rustlers of Devil's Canyon; Marshal of Cripple Creek; Along the Oregon Trail; Under Colorado Skies
1948: The Main Street Kid; Heart of Virginia; Secret Service Investigator; Out of the Storm; Son of God's Country; Sundown in Santa Fe; Renegades of Sonora
1949: Sheriff of Wichita; Death Valley Gunfighter; The Red Menace; Hellfire; Flame of Youth; Navajo Trail Raiders
1950: Belle of Old Mexico; Singing Guns; Harbor of Missing Men; The Arizona Cowboy; Hills of Oklahoma; Covered Wagon Raid; Frisco Tornado
1951: Million Dollar Pursuit; Honeychile; Street Bandits
1952: The Fabulous Senorita; Oklahoma Annie; Gobs and Gals; Tropical Heat Wave; Toughest Man in Arizona
1953: A Perilous Journey; Geraldine
1955: I Cover the Underworld; Double Jeopardy; Cross Channel (Brit.); Secret Venture (Brit.)
1956: Track the Man Down (Brit.); Come Next Spring; When Gangland Strikes
1957: Affair in Reno
1958: Cole Younger, Gunfighter; Revolt in the Big House
1959: Battle Flame; King of the Wild Stallions
1961: Operation Eichmann
1963: Showdown
1964: He Rides Tall; Bullet for a Badman; Taggart
1965: Black Spurs
1966: Apache Uprising; Johnny Reno; Waco
1967: Red Tomahawk; Hostile Guns
1968: Tiger by the Tail

JOHN M. STAHL (1886-1950)

1930: A Lady Surrenders
1931: Seed; Strictly Dishonorable
1932: Back Street
1933: Only Yesterday
1934: Imitation of Life
1935: Magnificent Obsession
1937: Parnell
1938: Letter of Introduction
1939: When Tomorrow Comes
1941: Our Wife
1942: The Immortal Sergeant

1943: Holy Matrimony
1944: The Eve of St. Mark; Keys of the Kingdom
1945: Leave Her to Heaven
1947: The Foxes of Harrow
1948: The Walls of Jericho
1949: Father Was a Fullback; Oh, You Beautiful Doll

SYLVESTER STALLONE (1946-)

1978: Paradise Alley
1979: Rocky II

JOHN STANLEY

1976: Nightmare in Blood

PAUL STANLEY

1959: Cry Tough
1968: Three Guns for Texas (w/ E. Bellamy and D. L. Rich)
1973: Cotter

JACK STARRETT

1969: Run, Angel, Run!
1970: The Losers
1972: Slaughter; The Strange Vengeance of Rosalio
1973: Cleopatra Jones
1974: The Gravy Train
1975: Race with the Devil
1976: A Small Town in Texas
1977: Final Chapter--Walking Tall

RALPH STAUB (1899-1969)

1936: Country Gentlemen; Sitting on the Moon
1937: Join the Marines; Navy Blues; The Mandarin Mystery; Affairs of Cappy Ricks; Meet the Boy Friend; Mama Runs Wild
1938: Prairie Moon; Western Jamboree
1940: Chip of the Flying U; Yukon Flight; Danger Ahead; Sky Bandits
1957: The Heart of Show Business

RAY DENNIS STECKLER (1939-)

1961: Wild Guitar

1963: Rat Fink a Boo Boo
1965: The Thrill Killers; The Incredibly-Strange Creatures Who Stopped Living and Became Mixed-Up Zombies
1974: The Chickenhawks

BOB STEIN

1970: The Zodiac Couples (w/ Alan Roberts)

JEFF STEIN

1979: The Kids Are Alright (doc.)

PAUL L. STEIN (1892-1951) Austrian

(American films only)
1929: This Thing Called Love; Her Private Affair; The Office Scandal
1930: The Lottery Bride; One Romantic Night; Sin Takes a Holiday
1931: Born to Love; The Common Law
1932: A Woman Commands; Lily Christine; Breach of Promise
1934: The Song You Gave Me

PETER STEPHENS

1959: Mustang

BERT STERN

1960: Jazz on a Summer's Day

LEONARD STERN

1979: Just You and Me, Kid

STEVEN HILLARD STERN

1971: P.S. I Love You
1974: Harrad Summer
1976: I Wonder Who's Killing Her Now?
1979: Running (Can.)

TOM STERN

1971: Clay Pigeon (w/ Lane Slate)

GEORGE STEVENS (1904-1975)

1933: The Cohens and Kellys in Trouble
1934: Bachelor Bait; Kentucky Kernels
1935: Alice Adams; Annie Oakley; The Nitwits; Laddie
1936: Swing Time
1937: A Damsel in Distress; Quality Street
1938: Vivacious Lady
1939: Gunga Din
1940: Vigil in the Night
1941: Penny Serenade
1942: The Talk of the Town; Woman of the Year
1943: The More the Merrier
1948: I Remember Mama
1951: A Place in the Sun
1952: Something to Live For
1953: Shane
1956: Giant
1959: The Diary of Anne Frank
1965: The Greatest Story Ever Told
1970: The Only Game in Town

LESLIE STEVENS (1924-)

1960: Private Property
1961: Incubus
1962: Hero's Island
1964: Della
1966: I Love a Mystery
1967: Fanfare for a Death Scene

MARK STEVENS (1915-)

1954: Cry Vengeance
1956: Timetable
1958: Gun Fever

ROBERT STEVENS (c. 1925-)

1957: The Big Caper
1958: Never Love a Stranger
1962: I Thank a Fool
1963: In the Cool of the Day
1969: Change of Mind

ROBERT STEVENSON (1905-) Brit.

(American films only)
1940: Tom Brown's School Days

1941: Back Street
1942: Joan of Paris
1943: Forever and a Day (w/ others)
1944: Jane Eyre
1947: Dishonored Lady
1948: To the Ends of the Earth
1949: I Married a Communist
1950: The Woman on Pier 13; Walk Softly, Stranger
1951: My Forbidden Past
1952: The Las Vegas Story
1957: Johnny Tremain; Old Yeller
1959: Darby O'Gill and the Little People
1960: Kidnapped
1961: The Absent-Minded Professor
1962: In Search of the Castaways
1963: Son of Flubber
1964: The Misadventures of Merlin Jones; Mary Poppins
1965: The Monkey's Uncle; That Darn Cat
1967: The Gnome-Mobile
1968: Blackbeard's Ghost
1969: The Love Bug
1971: Bedknobs and Broomsticks
1974: Herbie Rides Again; The Island at the Top of the World
1976: One of Our Dinosaurs Is Missing; The Shaggy D. A.

PETER STEWART (see SAM NEWFIELD)

JOHN STIX

1959: The Great St. Louis Bank Robbery (w/ C. Guggenheim)

BENJAMIN STOLOFF (1895-)

1929: Speakeasy; The Girl from Havana; Protection
1930: Happy Days; New Movietone Follies of 1930; Soup to Nuts
1931: Three Rogues; Goldie; Not Exactly Gentlemen
1932: The Night Mayor; The Devil Is Driving; Destry Rides Again; By Whose Hands?
1933: Obey the Law; Night of Terror
1934: Palooka; Transatlantic Merry-Go-Round
1935: To Beat the Band; Swell Head
1936: Sea Devils; Two in the Dark; Don't Turn 'Em Loose
1937: Super-Sleuth; Fight for Your Lady
1938: Radio City Revels; The Affairs of Annabel
1939: The Lady and the Mob
1940: The Marines Fly High (w/ G. Nicholls, Jr.)
1941: The Great Mr. Nobody; Three Sons o' Guns
1942: The Hidden Hand; Secret Enemies

1943: The Mysterious Doctor
1944: Take It or Leave It; Bermuda Mystery
1946: Johnny Comes Flying Home
1947: It's a Joke, Son

VICTOR STOLOFF

1966: Intimacy
1971: Three Hundred Year Weekend

ANDREW L. STONE (1902-)

1932: Hell's Headquarters
1937: The Girl Said No
1938: Say It in French; Stolen Heaven
1939: The Great Victor Herbert
1941: There's Magic in Music; The Hard-Boiled Canary
1943: Hi Diddle Diddle; Stormy Weather
1944: Sensations of 1945
1945: Bedside Manner
1946: The Bachelor's Daughters
1947: Fun on a Weekend
1950: Highway 301
1952: The Steel Trap; Confidence Girl
1953: A Blueprint for Murder
1955: The Night Holds Terror
1956: Julie
1958: Cry Terror; The Decks Ran Red
1960: The Last Voyage
1961: Ring of Fire
1963: The Password Is Courage (Brit.)
1964: Never Put It in Writing
1965: The Secret of My Success (Brit.)
1970: Song of Norway
1972: The Great Waltz (Brit.)

B. ZIGGY STONE

1971: Kovacs (doc.)

EZRA STONE

1967: Tammy and the Millionaire (w/ others)

MARSHALL STONE

1967: Come Spy with Me

OLIVER STONE

1974: Seizure

PHIL STONE

1937: Damaged Goods

VIRGINIA STONE

1976: Evil in the Deep

JEROME STORM

1929: Courtin' Wildcats
1932: The Racing Strain

LESLIE STORM

1946: Great Day

LOUIS CLYDE STOUMEN

1957: The Naked Eye (doc.)
1962: The Black Fox (doc.)
1965: The Image of Love

PAUL STRAND (1890-1976)

1942: Native Land (doc. w/ Leo Hurwitz)

FRANK R. STRAYER (1891-1964)

1929: The Fall of Eve; Acquitted
1930: Let's Go Places; Borrowed Wives
1931: Murder at Midnight; Caught Cheating; Anybody's Blonde; Soul of the Slums
1932: The Crusader; Dragnet Patrol; The Monster Walks; Behind Stone Walls; Love in High Gear; Gorilla Ship; Dynamite Denny; Tangled Destinies; Manhattan Tower
1933: Dance, Girl, Dance; The Vampire Bat; By Appointment Only
1934: In Love with Life; Fugitive Road; In the Money; Twin Husbands; Cross Streets; Fifteen Wives; One in a Million
1935: The Ghost Walks; Death from a Distance; Port of Lost Dreams; Symphony of Living; Public Opinion; Society Fever
1936: Hitch Hike to Heaven; Murder at Glen Athol; Sea Spoilers

1937: Off to the Races; Big Business; Hot Water; Borrowing Trouble; Laughing at Trouble
1938: Blondie
1939: Blondie Meets the Boss; Blondie Brings Up Baby; Blondie Takes a Vacation
1940: Blondie on a Budget; Blondie Has Servant Trouble; Blondie Plays Cupid
1941: Blondie Goes Latin; Go West, Young Lady; Blondie in Society
1942: Blondie's Blessed Event; Blondie Goes to College; Blondie for Victory
1943: The Daring Young Man; Footlight Glamour; It's a Great Life
1945: Mama Loves Papa; Senorita from the West
1946: I Ring Doorbells
1950: Messenger of Peace
1951: The Sickle and the Cross

JOSEPH STRICK (1923-)

1948: Muscle Beach (doc. w/ Irving Lerner)
1953: The Big Break (doc.)
1959: The Savage Eye (doc. w/ others)
1962: The Balcony
1967: Ulysses
1970: Tropic of Cancer
1974: Road Movie
1979: A Portrait of the Artist as a Young Man

HERBERT L. STROCK (1918-)

1954: Gog
1955: Battle Taxi
1957: Blood of Dracula; I Was a Teen-age Frankenstein
1958: How to Make a Monster
1962: Rider on a Dead Horse; The Devil's Messenger
1964: The Crawling Hand
1973: Brother on the Run

WILLIAM R. STROMBERG

1977: The Crater Lake Monster

ALLEN STUART

1938: Unashamed

MEL STUART

1964: Four Days in November (doc.)

1969: If It's Tuesday, This Must Be Belgium
1970: I Love My Wife
1971: Willy Wonka and the Chocolate Factory (Brit.)
1972: One Is a Lonely Number
1973: Wattstax (doc.); Visions of Eight (w/ others; Germ.)
1978: Mean Dog Blues

JOHN STUMAR

1937: The King's People

JOHN STURGES (1911-   )

1946: The Man Who Dared; Shadowed
1947: Alias Mr. Twilight; For the Love of Rusty; Keeper of the Bees; Thunderbolt (w/ W. Wyler)
1948: Sign of the Ram; Best Man Wins
1949: The Walking Hills
1950: The Capture; The Magnificent Yankee; Mystery Street; Right Cross
1951: Kind Lady; The People Against O'Hara; It's a Big Country (w/ others)
1952: The Girl in White
1953: Escape from Fort Brava; Fast Company; Jeopardy
1954: Bad Day at Black Rock
1955: The Scarlet Coat; Underwater!
1956: Backlash
1957: Gunfight at the O.K. Corral
1958: The Law and Jack Wade; The Old Man and the Sea (w/ others)
1959: Last Train from Gun Hill; Never so Few
1960: The Magnificent Seven
1961: By Love Possessed
1962: A Girl Named Tamiko; Sergeants 3
1963: The Great Escape
1965: The Hallelujah Trail; The Satan Bug
1967: Hour of the Gun
1968: Ice Station Zebra
1969: Marooned
1972: Joe Kidd
1974: McQ; Valdez the Halfbreed (Ital.)
1977: The Eagle Has Landed

PRESTON STURGES (1898-1959)

1940: Christmas in July; The Great McGinty
1941: The Lady Eve; Sullivan's Travels
1942: The Palm Beach Story
1944: The Great Moment; Hail the Conquering Hero; The Miracle at Morgan's Creek

1947: The Sin of Harold Diddlebock
1948: Unfaithfully Yours
1949: The Beautiful Blonde from Bashful Bend
1950: Mad Wednesday
1957: The French They Are a Funny Race (Brit.)

RONALD SULLIVAN

1970: Scorpio 70

HENRY SUSO

1978: Deathsport (w/ Allan Arkush)

A. EDWARD SUTHERLAND (1895-1974)

1929: Fast Company; The Saturday Night Kid; Pointed Heels
1930: Burning Up; The Social Lion; The Sap from Syracuse; Paramount on Parade (w/ others)
1931: Palmy Days; The Gang Buster; June Moon; Up Pops the Devil
1932: Mr. Robinson Crusoe; Secrets of the French Police; Sky Devils
1933: International House; Murder in the Zoo; Too Much Harmony
1935: Diamond Jim; Mississippi
1936: Poppy
1937: Champagne Waltz; Every Day's a Holiday
1939: The Flying Deuces
1940: Beyond Tomorrow; The Boys from Syracuse; One Night in the Tropics
1941: The Invisible Woman; Steel Against the Sky; Nine Lives Are Not Enough
1942: The Navy Comes Through; Sing Your Worries Away; Army Surgeon
1943: Dixie
1944: Follow the Boys; Secret Command
1945: Having Wonderful Crime
1946: Abie's Irish Rose

HAL SUTHERLAND

1974: Journey Back to Oz (anim.)

E. W. SWACKHAMER

1972: Man and Boy

CHARLES (CHUCK) SWENSON

1977: Dirty Duck (anim.)
1978: The Mouse and His Child (anim. w/ Fred Wolf)

DAVID SWIFT (1919-)

1960: Pollyanna
1961: The Parent Trap
1962: The Interns
1963: Love Is a Ball; Under the Yum Yum Tree
1964: Good Neighbor Sam
1967: How to Succeed in Business Without Really Trying

SAUL SWIMMER

1972: The Concert for Bangladesh (doc.)

BRAD SWIRNOFF

1976: Tunnelvision

PETER SYKES

1979: Jesus (w/ John Kirsch)

PAUL SYLBERT

1971: The Steagle

JEANNOT SZWAROC

1973: Extreme Close-Up
1975: Bug
1978: Jaws 2

ERROL TAGGART

1936: Women Are Trouble; The Longest Night; Sinner Take All
1937: Song of the City; Women Men Marry
1938: Strange Faces

GREGG TALLAS

1948: Siren of Atlantis
1951: Prehistoric Women
1956: Barefoot Battalion
1957: Bed of Grass
1967: Bikini Paradise

RICHARD TALMADGE (1896-1981)

1950: Border Outlaws
1953: Project Moonbase
1956: I Killed Wild Bill Hickok

(Serial)

1932: The Devil Horse (w/ O. Brower)

HARRY TAMPA

1979: Fairy Tales; Nocturna--Granddaughter of Dracula

JAMES TANSEY

1930: Romance of the West (w/ Robert Tansey)

ROBERT E. TANSEY (-1951)

1930: Romance of the West (w/ James Tansey)
1931: Riders of Rio
1932: The Galloping Kid
1941: The Driftin' Kid; Dynamite Canyon; Ridin' the Sunset Trail; Lone Star Lawmen
1942: Texas to Bataan; Trail Riders; Two Fisted Justice; Arizona Round-Up; Where Trails End; Western Mail
1943: Blazing Guns; Death Valley Rangers
1944: Arizona Whirlwind; Outlaw Trail; Sonora Stagecoach; Westward Bound; Harmony Trail
1945: Song of Old Wyoming
1946: God's Country; Colorado Serenade; Driftin' River; Stars over Texas; Tumbleweed Trail; Wild West; The Caravan Trail; Romance of the West
1948: The Enchanted Valley; Prairie Outlaws; Shaggy
1950: The Fighting Stallion; Forbidden Jungle; Federal Man
1951: Badman's Gold

DANIEL TARADASH

1956: Storm Center

FRANK TASHLIN (1913-1972)

1951: The Lemon Drop Kid (unc. w/ S. Lanfield)
1952: The First Time; Son of Paleface
1953: Marry Me Again
1954: Susan Slept Here
1955: Artists and Models
1956: The Girl Can't Help It; Hollywood or Bust; The Lieutenant Wore Skirts
1957: Will Success Spoil Rock Hunter?
1958: The Geisha Boy; Rock-a-Bye Baby
1959: Say One for Me
1960: Cinderfella
1961: Bachelor Flat
1962: It's Only Money
1963: The Man from the Diner's Club; Who's Minding the Store?
1964: The Disorderly Orderly
1966: The Glass Bottom Boat; The Alphabet Murders (Brit.)
1967: Caprice
1968: The Private Navy of Sgt. O'Farrell

NORMAN TAUROG (1899-1981)

1929: Lucky Boy (w/ Charles C. Wilson)
1930: Troopers Three (w/ B. R. Eason); Sunny Skies; Hot Curves; Follow the Leader
1931: Finn and Hattie (w/ N. Z. McLeod); Skippy; Newly Rich (alt. Forbidden Adventures); Huckleberry Finn; Sooky
1932: Hold 'Em Jail!; The Phantom President; If I Had a Million (w/ others)
1933: A Bedtime Story; The Way to Love
1934: We're Not Dressing; Mrs. Wiggs of the Cabbage Patch; College Rhythm
1935: The Big Broadcast of 1936
1936: Strike Me Pink; Rhythm on the Range; Reunion
1937: Fifty Roads to Town; You Can't Have Everything
1938: The Adventures of Tom Sawyer; Mad About Music; Boys Town
1939: The Girl Downstairs; Lucky Night
1940: Young Tom Edison; Broadway Melody of 1940; Little Nellie Kelly
1941: Men of Boys Town; Design for Scandal
1942: Are Husbands Necessary? A Yank at Eton
1943: Presenting Lily Mars; Girl Crazy
1946: The Hoodlum Saint
1947: The Beginning or the End
1948: The Bride Goes Wild; The Big City; Words and Music
1949: That Midnight Kiss
1950: Please Believe Me; The Toast of New Orleans; Mrs. O'Malley and Mr. Malone
1951: Rich, Young and Pretty

1952: Room for One More; Jumping Jacks; The Stooge
1953: The Stars Are Singing; The Caddy
1954: Living It Up
1955: You're Never Too Young
1956: The Birds and the Bees; Pardners; Bundle of Joy
1957: The Fuzzy Pink Nightgown
1958: Onionhead
1959: Don't Give up the Ship
1960: Visit to a Small Planet; G.I. Blues
1961: All Hands on Deck; Blue Hawaii
1962: Girls ! Girls! Girls!
1963: It Happened at the World's Fair; Palm Springs Weekend
1965: Tickle Me; Sergeant Deadhead; Dr. Goldfoot and the Bikini Machine
1966: Spinout
1967: Double Trouble
1968: Speedway; Live a Little, Love a Little

DON TAYLOR (1920-)

1961: Everything's Ducky
1964: Ride the Wild Surf
1967: Jack of Diamonds; The Five Man Army (Ital./Span.)
1971: Escape from the Planet of the Apes
1973: Tom Sawyer
1976: Echoes of a Summer; The Great Scout and Cathouse Thursday
1977: The Island of Dr. Moreau
1978: Damien--Omen II

RAY TAYLOR (1888?-1952)

1929: Come Across
1931: The One Way Trail
1934: Return of Chandu; Fighting Trooper
1935: The Ivory-Handled Gun; Outlawed Guns; The Throwback; Fang and Claw; Sunset of Power; Call of the Savage
1936: The Cowboy and the Kid; Silver Spurs; The Three Mesquiteers; Tex Rides with the Boy Scouts
1937: The Mystery of the Hooded Horseman; Sudden Bill Dorn; Drums of Destiny; Boss of Lonely Valley
1938: Frontier Town; Rawhide; Hawaiian Buckaroo; Panamint's Bad Man
1940: West of Carson City; Bad Man from Red Butte; Law and Order; Pony Post; Ragtime Cowboy Joe; Riders of Pasco Basin
1941: Fighting Bill Fargo; Boss of Bullion City; Bury Me Not on the Lone Prairie; Law of the Range; The Man from Montana; Rawhide Rangers; Lucky Ralston
1942: Destination Unknown; Treat 'Em Rough; Mountain Justice; Stagecoach Buckaroo

1943: Mug Town; Cheyenne Round-Up; The Lone Star Trail
1944: Boss of Boomtown
1945: The Daltons Ride Again
1946: The Michigan Kid
1947: The Vigilantes Return; Range Beyond the Blue; Border Feud; Law of the Lash; Wild Country; West to Glory; Ghost Town Renegades; Pioneer Justice; Black Hills; Return of the Lash; The Fighting Vigilantes; Stage to Mesa City; Cheyenne Takes Over
1948: Gunning for Justice; Hidden Danger; The Return of Wildfire; The Hawk of Powder River; Tornado Range; The Tioga Kid; The Westland Trail; Check Your Guns
1949: Law of the West; Range Justice; West of El Dorado; Son of Billy the Kid; Son of a Badman; Outlaw Country; Crashing Thru; Shadows of the West

(Serials)

1930: The Jade Box
1931: Battling with Buffalo Bill; Danger Island; Finger Prints; Detective Lloyd
1932: The Air Mail Mystery; Heroes of the West; The Jungle Mystery
1933: Clancy of the Mounted; Gordon of Ghost City; The Phantom of the Air
1934: The Pirate Treasure; The Return of Chandu
1935: The Roaring West; Tailspin Tommy and the Great Air Mystery
1936: The Phantom Rider; Robinson Crusoe on Clipper Island (w/ M. V. Wright); The Vigilantes Are Coming (w/ M. V. Wright)
1937: Dick Tracy (w/ Alan James); The Painted Stallion (w/ W. Witney); Flaming Frontier (w/ A. James)
1938: The Spider's Web (w/ J. Horne)
1939: Flying G-Men (w/ J. Horne); Scouts to the Rescue (w/ A. James); The Green Hornet
1940: Flash Gordon Conquers the Universe; Winners of the West (w/ F. Beebe)
1941: Riders of Death Valley; Sky Raiders (both w/ F. Beebe)
1942: Don Winslow of the Navy (w/ F. Beebe); Gang Busters (w/ Noel Smith); Junior G-Men of the Air (w/ L. Collins)
1943: Adventures of the Flying Cadets; The Adventures of Smilin' Jack; Don Winslow of the Coast Guard (all w/ L. Collins)
1944: The Great Alaskan Mystery; Mystery of the River Boat; Raiders of Ghost City (all w/ L. Collins)
1945: Jungle Queen; The Master Key; The Royal Mounted Rides Again; Secret Agent X-9 (all w/ L. Collins)
1946: Lost City of the Jungle; The Scarlet Horseman (both w/ L. Collins)

RICHARD TAYLOR

1978: Stingray

SAM TAYLOR (1895-1958)

1929: The Taming of the Shrew; Coquette
1930: DuBarry, Woman of Passion
1931: Kiki; Skyline; Ambassador Bill
1932: Devil's Lottery
1933: Out All Night
1934: The Cat's Paw
1935: Vagabond Lady
1944: Nothing but Trouble

SAMUEL A. TAYLOR

1957: The Monte Carlo Story

LEWIS TEAGUE

1974: Dirty O'Neil (w/ Howard Freen)
1979: The Lady in Red

GEORGE TEMPLETON

1950: The Sundowners
1951: Quebec

DEL TENNEY

1964: The Horror of Party Beach; The Curse of the Living Corpse

BERT TENZER

1969: 2,000 Years Later

DUCCIO TESSARI

1974: Three Tough Guys

TED TETZLAFF (1903-)

1941: World Premiere
1947: Riffraff
1948: Fighting Father Dunne
1949: The Window; Johnny Allegro; A Dangerous Profession
1950: The White Tower; Under the Gun; Gambling House
1952: The Treasure of Lost Canyon
1953: Terror on a Train
1955: Son of Sinbad

1956: Seven Wonders of the World (doc. w/ others)
1959: The Young Land

JOAN TEWKESBURY

1979: Old Boyfriends

PETER TEWKSBURY (1924-)

1963: Sunday in New York
1964: Emil and the Detectives (Germ.)
1967: Doctor, You've Got to Be Kidding!
1968: Stay Away, Joe
1969: The Trouble with Girls

WILLIAM THIELE (1890-1975) Austrian

(American films only)
1935: Lottery Lover
1936: The Jungle Princess
1937: London by Night; Beg, Borrow or Steal
1939: Bridal Suite; Bad Little Angel
1940: The Ghost Comes Home
1943: Tarzan Triumphs; Tarzan's Desert Mystery
1946: The Madonna's Secret

ROBERT THOM

1969: Angel, Angel, Down We Go (alt. Cult of the Damned)

MICHAEL THOMAS (Brit.)

(American films only)
1949: Once Upon a Dream
1952: Clouded Yellow; Island Rescue
1968: The High Commissioner

WILLIAM C. THOMAS (1903-)

1945: Midnight Manhunt; One Fascinating Night
1946: They Made Me a Killer; I Cover Big Town
1947: Big Town; Big Town After Dark (ret. Underworld After Dark)
1948: Big Town Scandal
1949: Special Agent

HARRY THOMASIN

1975: The Great Lester Boggs; So Sad About Gloria; Encounter with the Unknown
1979: The Day It Came to Earth

HARLAN THOMPSON

1933: The Past of Mary Holmes
1934: Kiss and Make Up

J. LEE THOMPSON (Brit.)

(American films only)
1960: I Aim at the Stars
1961: The Guns of Navarone
1962: Cape Fear; Taras Bulba
1963: Kings of the Sun
1964: John Goldfarb, Please Come Home; What a Way to Go!
1967: Eye of the Devil
1968: Before Winter Comes
1969: The Chairman; McKenna's Gold
1970: Brotherly Love
1972: Conquest of the Planet of the Apes
1973: Battle for the Planet of the Apes
1974: Huckleberry Finn
1975: The Reincarnation of Peter Proud
1976: St. Ives
1978: The Greek Tycoon

MARSHALL THOMPSON

1964: A Yank in Viet Nam

WALTER THOMPSON

1956: Seven Wonders of the World (doc. w/ others)
1958: South Sea Adventure (doc. w/ others)

WILLIAM L. THOMPSON

1935: The Irish Gringo

JERRY THORPE

1967: The Venetian Affair

1968: Day of the Evil Gun
1972: Company of Killers

RICHARD THORPE (1896-)

1929: The Bachelor Girl
1930: Border Romance; The Dude Wrangler; Wings of Adventure; The Thoroughbred; Under Montana Skies; The Utah Kid
1931: The Lawless Woman; The Lady from Nowhere; Wild Horse (w/ Sidney Algier); The Sky Spider; Grief Street; Neck and Neck; The Devil Plays
1932: Cross Examination; Forgotten Women; Murder at Dawn; Probation; Midnight Lady; Escapade; Forbidden Company; The Beauty Parlor; The King Murder; Thrill of Youth; Slightly Married
1933: Women Won't Tell; The Secrets of Wu Sin; Love Is Dangerous; Forgotten; Strange People; I Have Lived; Notorious but Nice; Man of Sentiment; Rainbow over Broadway
1934: Murder on the Campus; The Quitter; City Park; Stolen Sweets; Green Eyes; Cheating Cheaters
1935: Secret of the Chateau; Strange Wives; Last of the Pagans
1936: The Voice of Bugle Ann; Tarzan Escapes
1937: Dangerous Number; Night Must Fall; Double Wedding
1938: Man-Proof; Love Is a Headache; The First Hundred Years; The Toy Wife; The Crowd Roars; Three Loves Has Nancy
1939: The Adventures of Huckleberry Finn; Tarzan Finds a Son!
1940: The Earl of Chicago; Twenty-Mule Team; Wyoming
1941: The Bad Man; Barnacle Bill; Tarzan's Secret Treasure
1942: Joe Smith, American; Tarzan's New York Adventure; Apache Trail; White Cargo
1943: Three Hearts for Julia; Above Suspicion; Cry Havoc
1944: Two Girls and a Sailor; The Thin Man Goes Home
1945: Thrill of a Romance; Her Highness and the Bellboy; What Next, Corporal Hargrove?
1947: Fiesta; This Time for Keeps
1948: On an Island with You; A Date with Judy
1949: The Sun Comes Up; Big Jack; Challenge to Lassie; Malaya
1950: The Black Hand; Three Little Words
1951: Vengeance Valley; The Great Caruso; The Unknown Man; It's a Big Country (w/ others)
1952: Carbine Williams; Ivanhoe; The Prisoner of Zenda
1953: The Girl Who Had Everything; All the Brothers Were Valiant; Knights of the Round Table
1954: The Student Prince; Athena
1955: The Prodigal; Quentin Durward
1957: Ten Thousand Bedrooms; Tip on a Dead Jockey; Jailhouse Rock
1959: The House of the Seven Hawks
1960: Killers of Kilimanjaro

1961: The Honeymoon Machine
1962: The Horizontal Lieutenant; The Tartars (Ital.)
1963: Follow the Boys; Fun in Acapulco
1965: The Truth About Spring; That Funny Feeling; The Golden Head (Hung./U.S.)
1967: The Last Challenge

(Serials)

1929: King of the Kongo
1930: The Lone Defender
1931: King of the Wild

ALEXIS THURN-TAXIS (1891-1979) Austrian

(American films only)
1942: A Night for Crime; The Yanks Are Coming
1943: Man of Courage
1945: Hollywood and Vine

ERIC TILL

1968: Hot Millions
1970: The Walking Stick
1979: All Things Bright and Beautiful

JAMES TINLING (1889-1955)

1929: True Heaven; The Exalted Flapper; Words and Music
1930: For the Love o' Lil
1931: The Flood
1933: Arizona to Broadway; The Last Trail; Jimmy and Sally
1934: Three on a Honeymoon; Call It Luck; Love Time
1935: Senora Casada Necesita Marido (Span.); Under the Pampas Moon; Welcome Home; Charlie Chan in Shanghai
1936: Every Saturday Night; Champagne Charlie; Educating Father; Pepper; Back to Nature
1937: The Holy Terror; Angel's Holiday; Sing and Be Happy; The Great Hospital Mystery; 45 Fathers
1938: Change of Heart; Mr. Moto's Gamble; Passport Husband; Sharpshooters
1939: Boy Friend
1941: Last of the Duanes; Riders of the Purple Sage
1942: Sundown Jim; The Lone Star Ranger
1943: Cosmo Jones--Crime Smasher
1946: Rendezvous 24; Deadline for Murder; Strange Journey; Dangerous Millions
1947: Second Chance; Roses Are Red
1948: Night Wind; Trouble Preferred
1951: Tales of Robin Hood

DIMITRI TIOMKIN (1899-)

1972: Tschaikovsky

JAMES TOBACK

1978: Fingers

THOMAS J. TOBIN

1977: Fraternity Row

NORMAN TOKAR (1920-)

1962: Big Red
1963: Savage Sam
1964: A Tiger Walks; Those Calloways
1966: The Ugly Dachshund; Follow Me, Boys!
1967: The Happiest Millionaire
1968: The Horse in the Gray Flannel Suit
1969: Rascal
1970: The Boatniks
1972: Snowball Express
1974: Where the Red Fern Grows
1975: Apple Dumpling Gang
1976: No Deposit, No Return
1977: Candleshoe
1978: The Cat from Outer Space

FRANCHOT TONE (1905-1968)

1958: Uncle Vanya

BURT TOPPER

1958: Hell Squad
1959: Tank Commandos; The Diary of a High School Bride
1964: War Is Hell; The Strangler
1969: The Devil's 8
1971: The Hard Ride
1976: The Day the Lord Got Busted

IVAN TORS (1916-)

1964: Rhino!
1965: Zebra in the Kitchen

ROBERT TOTTEN

1969: Death of a Gunfighter (w/ Don Siegel)
1971: The Wild Country

JACQUES TOURNEUR (1904-   ) Fr.

(American films only)
1939: They All Come Out; Nick Carter, Master Detective
1940: Phantom Raiders
1941: Doctors Don't Tell
1942: The Cat People
1943: I Walked with a Zombie; The Leopard Man
1944: Days of Glory; Experiment Perilous
1946: Canyon Passage
1947: Out of the Past
1948: Berlin Express
1949: Easy Living
1950: Stars in My Crown; The Flame and the Arrow
1951: Circle of Danger; Anne of the Indies
1952: Way of a Gaucho
1953: Appointment in Honduras
1955: Stranger on Horseback; Wichita
1956: Great Day in the Morning; Nightfall
1958: Curse of the Demon (Brit.); The Fearmakers
1959: Timbuktu
1960: The Giant of Marathon (Ital.)
1963: The Comedy of Terrors
1965: War Gods of the Deep (Brit.)

JACK TOWNLEY

1939: Home on the Prairie; Ridin' the Range
1941: The Pittsburgh Kid

JAMES TOWNLEY

1934: Guilty Parents

BUD TOWNSEND

1969: Nightmare in Wax
1972: The Folks at Red Wolf Inn
1976: Terror House

SHEPARD TRAUBE

1940: Street of Memories
1941: The Bride Wore Crutches; For Beauty's Sake

LUIS TRENKER

1933: The Rebel (w/ E. A. Knopf)
1939: The Challenge

JOHN TRENT

1970: Homer

SIMON TREVOR

1972: The African Elephant (doc.)

CHARLES TRIESCHMANN

1974: Two

GUS TRIKONIS

1976: Nashville Girl; The Student Body
1977: New Girl in Town; Moonshine County Express
1978: The Evil

JAN TROELL (1931-) Swed.

(American films only)
1974: Zandy's Bride
1979: Hurricane

DALTON TRUMBO (1905-1976)

1971: Johnny Got His Gun

DOUGLAS TRUMBULL

1972: Silent Running

GLENN TRYON (1894-1970)

1934: Gridiron Flash
1936: Two in Revolt; Easy to Take
1937: Small Town Boy
1938: The Law West of Tombstone
1939: Beauty for the Asking
1941: Double Date

1943: That Nazty Nuisance
1944: Meet Miss Bobby Socks
1949: Miss Mink of 1949

MICHAEL TUCHNER

1973: Fear Is the Key

WANDA TUCHOCK

1934: Finishing School (w/ George Nicholls, Jr.)

PHIL TUCKER

1953: Robot Master

LAWRENCE TURMAN

1971: The Marriage of a Young Stockbroker

DEAN TURNER

1973: Valley of Blood

FRANK TUTTLE (1892-1963)

1929: Marquis Preferred; The Studio Murder Mystery; The Greene Murder Case; Sweetie; Men Are Like That
1930: Only the Brave; The Benson Murder Case; Paramount on Parade (w/ others); True to the Navy; Love Among the Millionaires; Her Wedding Night
1931: No Limit; It Pays to Advertise; Dude Ranch
1932: This Reckless Age; This Is the Night; The Big Broadcast
1933: Dangerously Yours; Pleasure Cruise; Roman Scandals
1934: Ladies Should Listen; Springtime for Henry; Here Is My Heart
1935: All the King's Horses; The Glass Key; Two for Tonight
1936: College Holiday
1937: Waikiki Wedding
1938: Doctor Rhythm
1939: Paris Honeymoon; I Stole a Million; Charlie McCarthy, Detective
1942: This Gun for Hire; Lucky Jordan
1943: Hostages
1944: The Hour Before the Dawn
1945: The Great John L; Don Juan Quilligan
1946: Suspense; Swell Guy

1950: Le Traque (w/ Boris Lewin; Fr.)
1951: The Magic Face
1955: Hell on Frisco Bay
1956: A Cry in the Night
1959: Island of Lost Women

DANIEL B. ULLMAN

1957: Badlands of Montana

EDGAR GEORGE ULMER (1904-1972) Austrian

1929: People on Sunday (w/ Robert Siodmak; Germ.)
1933: Mister Broadway; Damaged Lives
1934: The Black Cat; Thunder over Texas
1935: From Nine to Nine
1937: Natalka Poltavka (Ukrainian); Green Fields (w/ Jacob Ben-Ami; Yiddish)
1938: The Singing Blacksmith (Yiddish)
1939: Cossacks Across the Danube (Ukrainian); The Light Ahead (Yiddish); The Marriage Broker (Yiddish); Moon over Harlem
1942: Tomorrow We Live
1943: My Son, the Hero; Girls in Chains; Isle of Forgotten Sins; Jive Junction
1944: Bluebeard
1945: Strange Illusion (alt. Out of the Night)
1946: Club Havana; Detour; The Wife of Monte Carlo; Her Sister's Secret; The Strange Woman
1947: Carnegie Hall
1948: Ruthless
1949: The Pirates of Capri (Ital.; Captain Sirocco)
1951: The Man from Planet X; St. Benny the Dip
1952: Babes in Bagdad
1955: Murder Is My Beat; The Naked Dawn
1957: Daughter of Dr. Jekyll; The Perjurer (Germ.)
1960: Hannibal (Ital.); The Amazing Transparent Man; Beyond the Time Barrier
1961: L'Atlantide (Ital.; U.S. title: Journey Beneath the Desert)
1965: The Cavern (Ital.)

PETER USTINOV (1921-)

1961: Romanoff and Juliet
1962: Billy Budd
1972: Hammersmith Is Out

ROGER VADIM (1928-   ) Fr.

(American films only)
1971: Pretty Maids All in a Row

A. H. VAN BUREN

1929: Hearts in Dixie (w/ P. H. Sloane)
1930: Prince of Diamonds (w/ Karl Brown)

DANIEL J. VANCE

1975: Trained to Kill

WILLIAM K. VANDERBILT

1933: Over the Seven Seas

CORTLANDT VAN DEUSEN

1932: Bachelor Mother

W. S. VAN DYKE II (1899-1943)

1929: The Pagan
1931: Trader Horn; Never the Twain Shall Meet; Guilty Hands; The Cuban Love Song
1932: Tarzan, the Ape Man; Night Court
1933: Penthouse; The Prizefighter and the Lady; Eskimo (doc.)
1934: Manhattan Melodrama; The Thin Man; Hide-Out; Forsaking All Others; Laughing Boy
1935: I Live My Life; Naughty Marietta
1936: After the Thin Man; The Devil Is a Sissy; His Brother's Wife; Love on the Run; Rose Marie; San Francisco
1937: Personal Property; They Gave Him a Gun; Rosalie
1938: Marie Antoinette; Sweethearts
1939: Andy Hardy Gets Spring Fever; Another Thin Man; It's a Wonderful World; Stand Up and Fight
1940: Bitter Sweet; I Love You Again; I Take This Woman
1941: Dr. Kildare's Victory; The Feminine Touch; Rage in Heaven; Shadow of the Thin Man
1942: I Married an Angel; Journey for Margaret; Cairo

MELVIN VAN PEEBLES (1932-   )

1968: The Story of a Three-Day Pass
1970: Watermelon Man

AGNES VARDA (1928-) Fr.

(American films only)
1969: Lions Love

JOHN VARLEY

1933: Enlighten Thy Daughter

MARCEL VARNEL (1894-1947) Fr.

(American films only)
1932: Silent Witness (w/ R. L. Hough); Chandu the Magician (w/ W. C. Menzies)
1933: Infernal Machine
1935: Dance Band
1936: All In; I Give My Heart; No Monkey Business; Public Nuisance No. 1
1938: The Loves of Madame DuBarry
1940: Let George Do It

BAYARD VEILLER

1929: The Trial of Mary Dugan

EDWARD D. VENTURINI

1938: In Old Mexico
1939: The Llano Kid

HENRI VERNEUIL

1959: Forbidden Fruit; Paris Hotel
1960: The Big Chief
1968: Guns for San Sebastian

STEPHEN F. VERONA

1974: The Lords of Flatbush (w/ Martin Davidson)
1976: Pipe Dreams
1979: Boardwalk

BRUNO VE SOTA

1956: Female Jungle

VICTOR VICAS

1955: No Way Back
1957: The Wayward Bus
1958: Count Five and Die

PHIL VICTOR

1957: My Gun Is Quick (w/ George White)

CHARLES VIDOR (1900-1959)

1934: Sensation Hunters; Double Door
1935: Strangers All; The Arizonian; His Family Tree
1936: Muss 'Em Up
1937: A Doctor's Diary; The Great Gambini; She's No Lady
1939: Romance of the Redwoods; Blind Alley; Those High Grey Walls
1941: New York Town; Ladies in Retirement
1942: The Tuttles of Tahiti
1943: The Desperadoes
1944: Cover Girl; Together Again
1945: A Song to Remember; Over 21
1946: Gilda
1948: The Loves of Carmen
1951: It's a Big Country (w/ others)
1952: Hans Christian Andersen
1953: Thunder in the East
1954: Rhapsody
1955: Love Me or Leave Me
1956: The Swan
1957: The Joker Is Wild; A Farewell to Arms (Ital.)
1960: Song Without End (finished by George Cukor; Austria)

KING VIDOR (1894-)

1929: Hallelujah!
1930: Billy the Kid; Not So Dumb
1931: The Champ; Street Scene
1932: Bird of Paradise; Cynara
1933: The Stranger's Return
1934: Our Daily Bread
1935: So Red the Rose; The Wedding Night
1936: The Texas Rangers
1937: Stella Dallas
1938: The Citadel (Brit.)
1940: Comrade X; Northwest Passage
1941: H. M. Pulham, Esq.
1944: An American Romance

1946: Duel in the Sun (w/ others)
1948: On Our Merry Way (orig. A Miracle Can Happen; w/ L. Fenton)
1949: Beyond the Forest; The Fountainhead
1951: Lightning Strikes Twice
1952: Japanese War Bride; Ruby Gentry
1955: Man Without a Star
1956: War and Peace
1959: Solomon and Sheba

BERTHOLD VIERTEL (1885-1953) Germ.

(American films only)
1929: Seven Faces; The One Woman Idea
1930: Man Trouble
1931: The Magnificent Lie; The Spy
1932: The Man from Yesterday; The Wiser Sex
1934: Little Friend (Brit.)
1936: Rhodes (Brit.); The Passing of the 3rd Floor Back (Brit.)

ROBERT G. VIGNOLA (1882-1953)

1933: Broken Dreams
1934: The Scarlet Letter
1935: The Perfect Clue
1937: The Girl from Scotland Yard

CHUCK VINCENT

1979: A Matter of Love; Summer Camp

ALBERT T. VIOLA

1970: Interplay

JOSEPH VIOLA

1971: Angels Hard as They Come
1974: The Hot Box

JOHN VITIOLI

1970: The Pleasure Game

VIRGIL VOGEL

1956: The Mole People

1957: The Kettles on Old MacDonald's Farm; The Land Unknown
1958: Terror in the Midnight Sun (Swed.)
1965: The Sword of Ali Baba

JOSEF VON STERNBERG (1894-1969) Germ.

(American films only)
1929: Thunderbolt
1930: Morocco
1931: Dishonored; An American Tragedy
1932: Blonde Venus; Shanghai Express
1934: The Scarlet Empress
1935: Crime and Punishment; The Devil Is a Woman
1936: The King Steps Out
1937: I, Claudius (unfinished)
1939: Sergeant Madden
1941: The Shanghai Gesture
1944: The Town
1946: Duel in the Sun (w/ others)
1952: Macao
1953: The Saga of Anathan (Jap.)
1957: Jet Pilot

ERICH VON STROHEIM (1885-1957)

1933: Walking Down Broadway (unreleased, revised by others, released under title Hello Sister!)

BERNARD VORHAUS (c. 1898-) Germ.

(American films only)
1937: A Smile in the Storm (w/ H. Sokal); Hideout in the Alps
1938: King of the Newsboys; Tenth Avenue Kid
1939: Meet Dr. Christian; Fisherman's Wharf; Way Down South
1940: The Courageous Dr. Christian; Three Faces West; The Refugee
1941: Lady from Louisiana; Mr. District Attorney in the Carter Case; Angels with Broken Wings; Hurricane Smith
1942: Affairs of Jimmy Valentine; Ice Capades Revue
1947: Bury Me Dead; Winter Wonderland
1948: The Amazing Mr. X (alt. The Spiritualist)
1950: So Young, So Bad
1951: Pardon My French

SLAVKO VORKAPICH (1892-1976) Yugo.

1931: I Take This Woman (w/ M. Gering)

MICHAEL WADLEIGH

1970: Woodstock (doc.)

GEORGE WAGGNER (1894-)

1938: Western Trails; Outlaw Express; Guilty Trails; Prairie Justice; Black Bandit; Ghost Town Riders
1939: The Mystery Plane; Wolf Call; Stunt Pilot; The Phantom Stage
1940: Drums of the Desert
1941: Man-Made Monster; Horror Island; South of Tahiti; Sealed Lips; The Wolf Man
1944: The Climax
1945: Frisco Sal; Shady Lady
1946: Tangier
1947: The Gunfighters
1949: The Fighting Kentuckian
1951: Operation Pacific
1957: Destination 80,000; Pawnee

JANE WAGNER

1978: Moment by Moment

GY WALDRON

1974: Moonrunners

HAL WALKER (1896-1972)

1945: Out of This World; Duffy's Tavern; Stork Club; Road to Utopia
1950: My Friend Irma Goes West; At War with the Army
1951: That's My Boy; Sailor Beware
1952: Road to Bali

JOHNNIE WALKER

1933: Mr. Broadway (doc.)

ROBERT WALKER

1958: Street of Darkness

STUART WALKER (1887-1941)

1931: The Secret Call

1932: The False Madonna; The Misleading Lady; Evenings for Sale
1933: Tonight Is Ours; The Eagle and the Hawk; White Woman
1934: Romance in the Rain; Great Expectations
1935: The Mystery of Edwin Drood; The Werewolf of London; Manhattan Moon

RICHARD WALLACE (1894-1951)

1929: Shopworn Angel; Innocents of Paris; River of Romance
1930: Seven Days' Leave (w/ John Cromwell); Anybody's War; The Right to Love
1931: Man of the World; Kick In; The Road to Reno
1932: Tomorrow and Tomorrow; Thunder Below
1933: The Masquerader
1934: Eight Girls in a Boat; The Little Minister
1936: Wedding Present
1937: John Meade's Woman; Blossoms on Broadway
1938: The Young in Heart
1939: The Under-Pup
1940: Captain Caution
1941: A Girl, a Guy and a Gob; She Knew All the Answers; Obliging Young Lady
1942: The Wife Takes a Flyer
1943: A Night to Remember; The Fallen Sparrow; My Kingdom for a Cook
1944: Bride by Mistake
1945: It's in the Bag; Kiss and Tell
1946: Because of Him
1947: Sinbad the Sailor; Framed; Tycoon
1948: Let's Live a Little
1949: Adventure in Baltimore; A Kiss for Corliss

RAOUL WALSH (1892-1980)

1929: The Cockeyed World; Hot for Paris
1930: The Big Trail
1931: The Man Who Came Back; Women of All Nations; The Yellow Ticket
1932: Me and My Gal; Wild Girl
1933: The Bowery; Going Hollywood; Sailor's Luck
1935: Under Pressure; Every Night at Eight; Baby Face Harrigan
1936: Big Brown Eyes; Klondike Annie; Spendthrift
1937: Artists and Models; Hitting a New High; You're in the Army Now (Brit.); When Thief Meets Thief (Brit.)
1938: College Swing
1939: The Roaring Twenties; St. Louis Blues
1940: Dark Command; They Drive by Night
1941: High Sierra; Manpower; The Strawberry Blonde; They Died with Their Boots On
1942: Desperate Journey; Gentleman Jim
1943: Background to Danger; Northern Pursuit

1944: Uncertain Glory
1945: The Horn Blows at Midnight; Objective Burma!; Salty O'Rourke
1946: The Man I Love
1947: Cheyenne; Pursued
1948: Fighter Squadron; One Sunday Afternoon; Silver River
1949: Colorado Territory; White Heat
1951: Along the Great Divide; Captain Horatio Hornblower; Distant Drums
1952: Blackbeard the Pirate; Glory Alley; The Lawless Breed; The World in His Arms
1953: Gun Fury; A Lion Is in the Streets; Sea Devils (Brit.)
1954: Saskatchewan
1955: Battle Cry; The Tall Men
1956: The King and Four Queens; The Revolt of Mamie Stover
1957: Band of Angels
1958: The Naked and the Dead
1959: A Private's Affair; The Sheriff of Fractured Jaw (Brit.)
1960: Esther and the King
1961: Marines, Let's Go!
1964: A Distant Trumpet

CHARLES WALTERS (1911-)

1947: Good News
1948: Easter Parade
1949: The Barkleys of Broadway
1950: Summer Stock
1951: Texas Carnival; Three Guys Named Mike
1952: The Belle of New York
1953: Dangerous When Wet; Easy to Love; Lili; Torch Song
1955: The Glass Slipper; The Tender Trap
1956: High Society
1957: Don't Go Near the Water
1959: Ask Any Girl
1960: Please Don't Eat the Daisies; Cimarron (w/ Anthony Mann)
1961: Two Lovers
1962: Billy Rose's Jumbo
1964: The Unsinkable Molly Brown
1966: Walk, Don't Run

FRED WALTON

1979: When a Stranger Calls

SAM WANAMAKER (1919-)

1969: The File of the Golden Goose (Brit.)
1970: The Executioner (Brit.)
1971: Catlow

1977: Sinbad and the Eye of the Tiger
1979: Charlie Muffin (Brit.)

CLYDE WARE

1971: No Drums, No Bugles

ANDY WARHOL (1928-   )

1966: The Chelsea Girls
1967: My Hustler; I, a Man; Bike Boy; The Nude Restaurant
1972: Heat
1973: L'Amour

JACK WARNER, JR. (1916-   )

1962: Brushfire!

JOHN WARNER (see EDGAR G. ULMER)

CHARLES MARQUIS WARREN (1912-   )

1951: Little Big Horn
1962: Hellgate
1953: Arrowhead; Flight to Tangier
1955: Seven Angry Men
1956: Tension at Table Rock
1957: The Black Whip; Trooper Hook; Back from the Dead; The Unknown Terror; Copper Sky; Ride a Violent Mile
1958: Cattle Empire; Blood Arrow; Desert Hell
1969: Charro!

JERRY WARREN

1958: The Incredible Petrified World
1963: Attack of the Mayan Mummy

MARK WARREN

1972: Come Back, Charleston Blue

FRED WARSHOFSKY

1975: The Outer Space Connection (doc.)

JOHN WATERS

1946: The Mighty McGurk

JOHN WATERS

1970: Mondo Trasho
1977: Desperate Living

WILLIAM WATSON

1939: Heroes in Blue

NATE WATT

1936: Hopalong Cassidy Returns; Navy Born; Trail Dust
1937: Borderland; Hills of Old Wyoming; North of the Rio Grande; Rustler's Valley; Carnival Queen
1939: Law of the Pampas
1940: Frontier Vengeance; Oklahoma Renegade

SAL WATTS

1974: Solomonking (w/ Jack Bomay)

JOHN WAYNE (1907-1979)

1960: The Alamo
1968: The Green Berets (w/ Ray Kellogg)

HARRY S. WEBB

1929: Dark Skies
1930: Bar L Ranch; Ridin' Law; Beyond the Rio Grande; Phantom of the Desert
1931: West of Cheyenne; Westward Bound
1932: Lone Trail
1933: Riot Squad; Wolf Rides
1934: Fighting Hero; Riding Thru; Tracy Rides; The Cactus Kid
1935: Unconquered Bandit
1936: Pinto Rustlers; Santa Fe Bound
1939: Port of Hate; Feud of the Range; Mesquite Buckaroo; Riders of the Sage; The Pal from Texas
1940: Pioneer Days

(Serial)

1931: The Sign of the Wolf (w/ F. Sheldon)

IRA WEBB

1939: El Diablo Rides

JACK WEBB (1920-   )

1954: Dragnet
1955: Pete Kelly's Blues
1957: The D.I.
1959: -30-
1961: The Last Time I Saw Archie

KENNETH WEBB

1929: Lucky in Love

MILLARD WEBB (1893-1935)

1929: Gentlemen of the Press; Glorifying the American Girl
1930: Her Golden Calf
1934: The Woman Who Dared

ROBERT D. WEBB (1903-   )

1945: The Caribbean Mystery; The Spider
1953: The Glory Brigade; Beneath the 12-Mile Reef
1955: White Feather; Seven Cities of Gold
1956: On the Threshold of Space; The Proud Ones; Love Me Tender
1957: The Way to the Gold
1960: Guns of the Timberland
1961: Pirates of Tortuga; Seven Women from Hell
1967: The Capetown Affair

LOIS WEBER (1882-1939)

1934: White Heat

NICHOLAS WEBSTER

1962: Dead to the World
1963: Gone Are the Days
1964: Santa Claus Conquers the Martians
1968: Mission Mars
1979: No Longer Alone

F. HARMON WEIGHT

1929: Frozen River; Hardboiled Rose

SAMUEL WEIL

1979: Squeeze Play

CLAUDIA WEILL (1947-)

1975: The Other Half of the Sky: A China Memoir (doc. w/ Shirley MacLaine)
1978: Girlfriends

LENNIE WEINRIB

1965: Beach Ball; Wild, Wild Winter
1966: Out of Sight

MARVIN WEINSTEIN

1956: Running Target

DON WEIS (1922-)

1951: Bannerline; It's a Big Country (w/ others)
1952: Just This Once; You for Me
1953: I Love Melvin; Remains to Be Seen; A Slight Case of Larceny; The Affairs of Dobie Gillis; Half a Hero
1954: The Adventures of Hajji Baba
1956: Ride the High Iron
1959: The Gene Krupa Story
1963: Critic's Choice
1964: Looking for Love; Pajama Party
1965: Billie
1966: The Ghost in the Invisible Bikini
1967: The King's Pirate
1968: Did You Hear the One About the Traveling Saleslady?

JACK WEIS

1972: Quadroon

ADRIAN WEISS

1953: The Bride and the Beast

BARBARA N. WEISS

1977: On the Line

MEL WELLES

1973: Lady Frankenstein

ORSON WELLES (1915-)

1941: Citizen Kane
1942: The Magnificent Ambersons; It's All True (unfinished)
1943: Journey into Fear (w/ N. Foster)
1946: The Stranger
1948: The Lady from Shanghai; Macbeth
1952: Othello (Ital.)
1955: Mr. Arkadin (Brit.; ret. Confidential Report)
1958: Touch of Evil
1959: Don Quixote (unfinished)
1963: The Trial (Fr./Ital./Germ.)
1966: Chimes at Midnight (alt. Falstaff)
1968: The Immortal Story (Fr.)

WILLIAM A. WELLMAN (1896-1975)

1929: Chinatown Nights; The Man I Love; Woman Trap
1930: Maybe It's Love; Dangerous Paradise; Young Eagles; Steel Highway
1931: Night Nurse; Other Men's Women; Public Enemy; Safe in Hell; Star Witness
1932: The Hatchet Man; Love Is a Racket; Purchase Price; The Conquerors; So Big
1933: College Coach; Heroes for Sale; Central Airport; Frisco Jenny; Lilly Turner; Wild Boys of the Road; Midnight Mary; Lady of the Night
1934: Looking for Trouble; Stingaree; The President Vanishes
1935: Call of the Wild; Robin Hood of El Dorado
1936: Small Town Girl
1937: Nothing Sacred; A Star Is Born
1938: Men with Wings
1939: Beau Geste; The Light That Failed
1940: Reaching for the Sun
1942: The Great Man's Lady; Roxie Hart: Thunder Birds
1943: Lady of Burlesque; The Ox-Bow Incident
1944: Buffalo Bill
1945: The Story of G.I. Joe; This Man's Navy
1946: Gallant Journey
1947: Magic Town
1948: The Happy Years; The Iron Curtain (ret. Behind the Iron Curtain); The Next Voice You Hear

1949: Battleground
1950: Across the Wide Missouri; It's a Big Country (w/ others); Westward the Women
1951: Yellow Sky
1952: My Man and I
1953: Island in the Sky
1954: The High and the Mighty; Track of the Cat
1955: Blood Alley
1956: Goodbye, My Lady
1958: Darby's Rangers; Lafayette Escadrille

WIM WENDERS Germ.

1977: The American Friend

PAUL WENDKOS (1922-)

1957: The Burglar
1958: The Case Against Brooklyn; Tarawa Beachhead
1959: Gidget; Face of a Fugitive; Battle of the Coral Sea
1960: Because They're Young
1961: Angel Baby (w/ H. Cornfield); Gidget Goes Hawaiian
1963: Gidget Goes to Rome
1966: Johnny Tiger
1968: Attack on the Iron Coast (Brit.)
1969: Guns of the Magnificent Seven
1970: Cannon for Cordoba (Span.)
1971: Hellboats; The Mephisto Waltz
1976: Special Delivery

ALFRED LOUIS WERKER (1896-)

1929: The Blue Skies; Chasing Through Europe (w/ David Butler)
1930: Double Cross Roads (w/ G. Middleton); The Last of the Duanes
1931: Fair Warning; Annabelle's Affairs; Heartbreak
1932: The Gay Caballero; Bachelor's Affairs; Rackety Rax
1933: It's Great to Be Alive; Advice to the Lovelorn
1934: The House of Rothschild; You Belong to Me
1935: Stolen Harmony
1936: Love in Exile (Brit.)
1937: We Have Our Moments; Wild and Woolly; City Girl; Big Town Girl
1938: Kidnapped; Gateway; Up the River
1939: It Could Happen to You; News Is Made at Night; The Adventures of Sherlock Holmes
1941: Moon over Her Shoulder; The Reluctant Dragon (w/ others)
1942: Whispering Ghosts; The Mad Martindales; A-Haunting We Will Go
1944: My Pal Wolf

1946: Shock
1947: Repeat Performance; Pirates of Monterey
1948: He Walked by Night
1949: Lost Boundaries
1951: Sealed Cargo
1952: Walk East on Beacon
1953: The Last Posse; Devil's Canyon
1954: Three Hours to Kill
1955: Canyon Crossroads; At Gunpoint
1956: Rebel in Town
1957: The Young Don't Cry

JEFF WERNER

1979: Cheerleaders' Wild Weekend

LINA WERTMULLER (1928-   ) Ital.

(American films only)
1977: The End of the World in Our Usual Bed in a Night Full of Rain

ROBERT D. WEST

1973: The Wednesday Children

ROLAND WEST (1887-1952)

1929: Alibi
1931: The Bat Whispers; Corsair

WILLIAM WEST

1940: The Last Alarm; Flying Wild

JIM WESTMAN

1974: The Wrestler

HASKELL WEXLER (1926-   )

1965: The Bus (doc.)
1969: Medium Cool

JAMES WHALE (1896-1957)

1930: Journey's End
1931: Waterloo Bridge; Frankenstein
1932: Impatient Maiden; The Old Dark House
1933: The Kiss Before the Mirror; The Invisible Man; By Candlelight
1934: One More River
1935: The Bride of Frankenstein; Remember Last Night?
1936: Showboat
1937: The Road Back; The Great Garrick
1938: Sinners in Paradise; Wives Under Suspicion; Port of Seven Seas
1939: The Man in the Iron Mask
1940: Green Hell
1941: They Dare Not Love
1952: Hello Out There (unreleased)

TIM WHELAN (1893-1957) Brit.

(American films only)
1935: The Murder Man; The Perfect Gentleman
1941: The Mad Doctor; International Lady
1942: Twin Beds; Nightmare; Seven Days' Leave
1943: Swing Fever; Higher and Higher
1944: Step Lively
1946: Badman's Territory
1955: Rage at Dawn; Texas Lady

GEORGE WHITE

1934: George White's Scandals
1957: My Gun Is Quick (w/ Phil Victor)

JULES WHITE

1931: Sidewalks of New York (w/ Z. Myers)
1960: Stop! Look! and Laugh!

MERRILL C. WHITE

1957: Ghost Diver

SAM WHITE

1941: The Officer and the Lady
1942: I Live on Danger
1945: People Are Funny

PHILLIP H. WHITMAN

- 1930: Fourth Alarm
- 1931: Mystery Train; Air Eagles
- 1932: Stowaway; The Girl from Calgary (w/ L. D'Usseau)
- 1933: Police Call; His Private Secretary; Strange Adventure

JOHN WHITMORE

- 1973: Here Comes Every Body (doc.)

RICHARD WHORF (1906-1966)

- 1944: Blonde Fever
- 1945: The Hidden Eye; The Sailor Takes a Wife
- 1946: Till the Clouds Roll By
- 1947: It Happened in Brooklyn; Love from a Stranger
- 1948: Luxury Liner
- 1950: Champagne for Caesar
- 1951: The Groom Wore Spurs

BERNHARD WICKI (1919-) Germ.

(American films only)
- 1960: The Bride
- 1962: The Longest Day (w/ others)
- 1964: The Visit
- 1965: Morituri

KEN WIEDERHORN

- 1979: King Frat

MICHAEL WIESE

- 1979: Dolphin (doc. w/ Hardy Jones)

JOE WIEZYCKI

- 1975: Satan's Children

ROBERT WILBOR

- 1976: Mark Twain, American

CRANE WILBUR (1889-1973)

1934: Tomorrow's Children
1935: High School Girl; The People's Enemy; The Rest Cure
1936: Devil on Horseback; Romance of Robert Burns; We're in the Legion Now; Yellow Cargo
1937: Navy Spy (w/ J. H. Lewis)
1938: The Patient in Room 18 (w/ Bobby Connolly)
1939: The Man Who Dared; I Am Not Afraid
1947: The Devil on Wheels
1948: Canon City
1949: The Story of Molly X
1950: Outside the Wall
1951: Inside the Walls of Folsom Prison
1959: The Bat

FRED McLEOD WILCOX (c. 1905-1964)

1943: Lassie Come Home
1946: Blue Sierra; Courage of Lassie
1948: Hills of Home; Three Daring Daughters
1949: The Secret Garden
1951: Shadow in the Sky
1953: Code Two
1954: Tennessee Champ
1956: Forbidden Planet
1960: I Passed for White

HERBERT WILCOX (1892-1977) Brit.

(American films only)
1939: Nurse Edith Cavell
1940: Irene; No, No, Nanette
1941: Sunny

CORNEL WILDE (1915-)

1955: Storm Fear
1957: The Devil's Hairpin
1958: Maracaibo
1963: Sword of Lancelot (Brit.)
1966: The Naked Prey
1967: Beach Red
1969: The Raging Sea
1970: No Blade of Grass
1975: Sharks' Treasure

TED WILDE

1930: Loose Ankles; Clancy in Wall Street

BILLY WILDER (1906-)

1942: The Major and the Minor
1943: Five Graves to Cairo
1944: Double Indemnity
1945: The Lost Weekend
1948: The Emperor Waltz; A Foreign Affair
1950: Sunset Boulevard
1951: The Big Carnival (orig. Ace in the Hole)
1953: Stalag 17
1954: Sabrina
1955: The Seven Year Itch
1957: Love in the Afternoon; The Spirit of St. Louis
1958: Witness for the Prosecution
1959: Some Like It Hot
1960: The Apartment
1961: One, Two, Three
1963: Irma la Douce
1964: Kiss Me, Stupid
1966: The Fortune Cookie
1970: The Private Life of Sherlock Holmes
1972: Avanti
1974: The Front Page
1979: Fedora

GENE WILDER (1935-)

1975: The Adventure of Sherlock Holmes' Smarter Brother
1977: The World's Greatest Lover

W. LEE WILDER

1946: The Glass Alibi
1947: The Pretender; Yankee Pasha
1948: The Vicious Circle
1950: Once a Thief
1951: Three Steps North (Ital.)
1953: Phantom from Space
1954: Killers from Space; The Snow Creature
1955: The Big Bluff
1956: Manfish
1958: Spy in the Sky
1960: Bluebeard's Ten Honeymoons

GORDON WILES

1936: Blackmailer; Charlie Chan's Secret; Two-Fisted Gentleman; Lady from Nowhere
1937: Women of Glamour; Venus Makes Trouble
1938: Prison Train; Mr. Boggs Steps Out
1941: Forced Landing

1947: The Gangster
1973: Ginger in the Morning

IRVIN WILLAT (1892-1976)

1931: Damaged Love
1937: Luck of Roaring Camp; Old Louisiana; South of Sonora; Under Strange Flags

PAUL B. WILLETT

1933: Home on the Range; Western Skies

ELMO WILLIAMS (1913-)

1953: The Tall Texan
1954: The Cowboy (doc.)
1957: Apache Warrior; Hell Ship Mutiny (w/ Lee Sholem)

LESTER WILLIAMS (see WILLIAM BERKE)

OSCAR WILLIAMS

1972: The Final Comedown
1973: Five on the Black Hand Side
1976: Hot Potato

PAUL WILLIAMS (1944-)

1969: Out of It
1970: The Revolutionary
1972: Dealings: Or the Berkeley to Boston Forty-Brick Lost Bag Blues
1978: Nunzio

RICHARD WILLIAMS

1977: Raggedy Ann and Andy (anim.)

SPENCER WILLIAMS

1941: Blood of Jesus
1944: Go Down Death
1945: The Girl in Room 20
1946: Beale Street Mama; Dirty Gertie from Harlem, U.S.A.
1947: Jivin' in Be Bop; Juke Joint

FRED WILLIAMSON (1938-)

1975: Adios Amigo
1976: Death Journey; Mean Johny Barrows; No Way Back

J. E. WILLIAMSON

1932: With Williamson Beneath the Sea (doc.)

JAY WILSEY (BUFFALO BILL, JR.)

1934: Riding Speed *
1935: Trails of Adventure

CHARLES C. WILSON

1929: Lucky Boy (w/ N. Taurog)

FRANK ARTHUR WILSON

1976: Blast

JAMES WILSON

1976: Death Riders
1979: Screams of a Winter Night

JOHN D. WILSON

1971: Shinbone Alley (anim.)

RICHARD WILSON (1915-)

1955: Man with the Gun
1957: The Big Boodle
1958: Raw Wind in Eden
1959: Al Capone
1960: Pay or Die
1963: Wall of Noise
1964: Invitation to a Gunfighter
1968: Three in the Attic

LAWRENCE C. WINDON

1931: Enemies of the Law

BRETAIGNE WINDUST (1906-1960)

1948: Winter Meeting; June Bride
1950: Perfect Strangers; Pretty Baby
1951: The Enforcer (finished by Raoul Walsh)
1952: Face to Face (w/ John Brahm)

HARRY S. WINER

1976: Legend of Bigfoot

ROBERT WINER

1974: The Devil's Triangle

WARD WING

1933: Samarang

MICHAEL WINNER (1935-)

1962: Play It Cool (Brit.)
1963: West 11 (Brit.)
1966: You Must Be Joking! (Brit.); The Girl Getters (Brit.)
1967: The Jokers (Brit.)
1968: I'll Never Forget What's 'Is Name (Brit.)
1971: Lawman
1972: The Mechanic; The Nightcomers (Brit.); Chato's Land
1973: Scorpio; The Stone Killer
1974: Death Wish
1976: Won Ton Ton, The Dog That Saved Hollywood
1977: The Sentinel
1978: The Big Sleep
1979: Firepower

SUSAN WINSLOW

1976: All This and World War II (doc.)

RON WINSTON

1966: Ambush Bay
1967: Banning
1968: Don't Just Stand There
1970: The Gamblers

S. K. WINSTON

1944: Adventure in Music (w/ others)

DAVID WINTERS

1979: Racquet

FRANK WISBAR (1899-1967) E. Prussian

(American films only)
1945: Strangler of the Swamp
1946: Devil Bat's Daughter; Lighthouse; Secrets of a Sorority Girl
1947: The Prairie
1948: The Mozart Story
1964: Commando

AUBREY WISBERG

1954: Dragon's Gold (w/ Jack Pollexfen)

ROBERT WISE (1914-)

1944: The Curse of the Cat People (w/ Gunther Fritsch); Mademoiselle Fifi
1945: The Body Snatcher; A Game of Death
1946: Criminal Court
1947: Born to Kill
1948: Blood on the Moon; Mystery in Mexico
1949: The Set-Up
1950: Three Secrets; Two Flags West
1951: The Day the Earth Stood Still; House on Telegraph Hill
1952: Something for the Birds; The Captive City
1953: The Desert Rats; Destination Gobi; So Big
1954: Executive Suite
1955: Helen of Troy
1956: Somebody Up There Likes Me; Tribute to a Bad Man
1957: This Could Be the Night; Until They Sail
1958: I Want to Live!; Run Silent, Run Deep
1959: Odds Against Tomorrow
1961: West Side Story (w/ Jerome Robbins)
1962: Two for the Seesaw
1963: The Haunting
1965: The Sound of Music
1966: The Sand Pebbles
1968: Star!
1971: The Andromeda Strain
1973: Two People

1975: The Hindenburg
1977: Audrey Rose
1979: Star Trek--the Motion Picture

FRED WISEMAN

1966: Titicut Follies (doc.)

CHET WITHEY

1928: The Bushranger

PAUL WITHINGTON

1931: The Blonde Captive (doc.)

WILLIAM WITNEY (1910-   )

1937: The Trigger Trio
1940: Heroes of the Saddle; Hi-Yo Silver! (w/ John English)
1942: Outlaws of Pine Ridge
1946: Roll on Texas Moon; Home in Oklahoma; Helldorado
1947: Apache Rose; Bells of San Angelo; Springtime in the Sierras; On the Old Spanish Trail
1948: The Gay Ranchero; Under California Stars; Eyes of Texas; Nighttime in Nevada; Grand Canyon Trail; The Far Frontier
1949: Susanna Pass; Down Dakota Way; The Golden Stallion
1050: Bells of Coronado; Twilight in the Sierras; Trigger, Jr.; Sunset in the West; North of the Great Divide; Trail of Robin Hood
1951: Spoilers of the Plains; Heart of the Rockies; In Old Amarillo; South of Caliente; Pals of the Golden West
1952: Colorado Sundown; The Last Musketeer; Border Saddlemates; Old Oklahoma Plains; The WAC from Walla Walla; South Pacific Trail
1953: Old Overland Trail; Down Laredo Way; Shadows of Tombstone; Iron Mountain Trail
1954: The Outcast
1955: Santa Fe Passage; City of Shadows; Headline Hunters; The Fighting Chance
1956: Stranger at My Door; A Strange Adventure
1957: Panama Sal
1958: Juvenile Jungle; The Cool and the Crazy; The Bonnie Parker Story; Young and Wild
1959: Paratroop Command
1960: Valley of the Redwoods; The Secret of the Purple Reef
1961: The Long Rope; Master of the World; The Cat Burglar
1964: Apache Rifles

1965: The Girls on the Beach; Arizona Raiders
1967: Forty Guns to Apache Pass
1973: I Escaped from Devil's Island
1975: Darktown Strutters

(Serials)

1937: The Painted Stallion (w/ Ray Taylor); S.O.S. Coast Guard (w/ Alan James); Zorro Rides Again (w/ John English)
1938: Dick Tracy Returns; The Fighting Devil Dogs; Hawk of the Wilderness; The Lone Ranger (all w/ J. English)
1939: Daredevils of the Red Circle; Dick Tracy's G-Men; The Lone Ranger Rides Again; Zorro's Fighting Legion (all w/ J. English)
1940: Adventures of Red Ryder; Drums of Fu Manchu; Hi-Yo Silver; King of the Royal Mounted; Mysterious Dr. Satan (all w/ J. English)
1941: Adventures of Captain Marvel; Dick Tracy vs. Crime, Inc.; Jungle Girl; King of the Texas Rangers (all w/ J. English); King of the Mounties
1943: G-Men vs. the Black Dragon
1946: The Crimson Ghost (w/ Fred Brannon)
1952: Dick Tracy vs. Phantom Empire (w/ J. English)

IRA WOHL

1979: Best Boy

FRED WOLF

1978: The Mouse and His Child (anim. w/ Chuck Swenson)

TOM WOLF

1970: Wilbur and the Baby Factory

P. J. WOLFSON

1939: Boy Slaves

LOUIS WOLHEIM

1931: Sin Ship

EDWARD WOOD, JR.

1953: Glen or Glenda? (alt. I Led Two Lives)

1956: Bride of the Monster
1959: Plan Nine from Outer Space; Night of the Ghouls (unreleased)

PETER WOOD

1970: In Search of Gregory

SAM WOOD (1883-1949)

1929: So This Is College; It's a Great Life
1930: Way of a Sailor; The Girl Said No; Sins of the Children; Richest Man in the World; Paid
1931: A Tailor-Made Man; The Man in Possession; New Adventures of Get-Rich-Quick Wallingford
1932: Prosperity; Huddle
1933: Hold Your Man; The Barbarian; Christopher Bean
1934: Stamboul Quest
1935: A Night at the Opera; Whipsaw; Let 'Em Have It
1936: The Unguarded Hour
1937: Navy Blue and Gold; Madame X; A Day at the Races
1938: Lord Jeff; Stablemates
1939: Goodbye, Mr. Chips (Brit.)
1940: Kitty Foyle; Our Town; Raffles; Rangers of Fortune
1941: The Devil and Miss Jones
1942: The Pride of the Yankees; King's Row
1943: For Whom the Bell Tolls
1944: Casanova Brown
1945: Guest Wife
1946: Heartbeat; Saratoga Trunk
1947: Ivy
1949: Ambush; The Stratton Story; Command Decision

HORACE AND STACY WOODARD

1937: Adventures of Chico (doc.)

BOB WOODBURN

1973: Little Laura and Big John (w/ Luke Moberly)

FRANK WOODRUFF

1940: Play Girl; Curtain Call; Cross Country Romance; Wildcat Bus
1941: Lady Scarface
1943: Cowboy in Manhattan; Two Senoritas from Chicago; Pistol Packin' Mama
1944: Lady, Let's Dance

JACK WOODS

1970: Equinox

CARL WORKMAN

1976: The Money

J. C. WORKS (see CHESTER ERSKINE)

DUKE WORNE

1929: Bride of the Desert; Handcuffed
1930: Midnight Special
1932: The Last Ride

DAVID WORTH

1979: Hollywood Knight

BASIL WRANGELL

1947: Philo Vance's Gamble
1958: Cinerama--South Seas Adventure

JOHN GRIFFITH WRAY

1929: A Most Immoral Lady; The Careless Age

MACK V. WRIGHT (1895-1963)

1932: Haunted Gold
1933: The Man from Monterey; Somewhere in Sonora
1935: Cappy Ricks Returns
1936: The Big Show; Comin' Round the Mountain; Roarin' Lead (w/ S. Newfield); The Singing Cowboy; Winds of the Wasteland
1937: Hit the Saddle; Rootin' Tootin' Rhythm; Riders of the Whistling Skull; Range Defenders
1940: The Man from Tascosa (alt. Wells Fargo Days)

(Serials)

1936: Robinson Crusoe on Clipper Island; The Vigilantes Are Coming (both w/ Ray Taylor)
1938: Great Adventures of Wild Bill Hickok (w/ Sam Nelson)
1942: The Sea Hound (w/ B. Eason)

PATRICK WRIGHT

1976: Hollywood High

RALPH WRIGHT

1957: Perri (w/ Paul Kenworthy, Jr.)

TENNY WRIGHT

1932: The Big Stampede
1933: The Telegraph Trail

DONALD WRYE

1979: Ice Castles

WILLIAM WYLER (1902-1981)

1929: The Shakedown; Hell's Heroes; The Love Trap
1930: The Storm
1932: A House Divided; Tom Brown of Culver
1933: Her First Mate; Counsellor-at-Law
1934: Glamour
1935: The Good Fairy; The Gay Deception
1936: Come and Get It (w/ Howard Hawks); Dodsworth; These Three
1937: Dead End
1938: Jezebel
1939: Wuthering Heights
1940: The Letter; The Westerner
1941: The Little Foxes
1942: Mrs. Miniver
1946: The Best Years of Our Lives
1949: The Heiress
1951: Detective Story
1952: Carrie
1953: Roman Holiday
1955: The Desperate Hours
1956: Friendly Persuasion
1958: The Big Country
1959: Ben-Hur
1962: The Children's Hour
1965: The Collector
1966: How to Steal a Million
1968: Funny Girl
1970: The Liberation of L. B. Jones

BOB WYNN

1971: The Resurrection of Zachary Wheeler

YABLO YABLONSKY

1971: B. J. Presents

JEAN YARBROUGH (1900-)

1940: The Devil Bat
1941: Caught in the Act; South of Panama; King of the Zombies; The Gang's All Here; Father Steps Out; Let's Go Collegiate; Top Sergeant Mulligan; City Limits
1942: Freckles Comes Home; Man from Headquarters; Law of the Jungle; So's Your Aunt Emma!; She's in the Army; Police Bullets; Criminal Investigator; Lure of the Islands; Silent Witness
1943: Follow the Band; Good Morning, Judge; Get Going; Hi' Ya, Sailor; So's Your Uncle
1944: Weekend Pass; Moon over Las Vegas; South of Dixie; In Society; Twilight on the Prairie
1945: Under Western Skies; Here Come the Co-Eds; The Naughty Nineties; On Stage Everybody
1946: She Wolf of London; House of Horrors; Inside Job; Cuban Pete; The Brute Man
1948: The Challenge; Shed No Tears; The Creeper; Triple Threat
1949: Henry, the Rainmaker; The Mutineers; Leave It to Henry; Angels in Disguise; Holiday in Havana; Master Minds
1950: Joe Palooka Meets Humphrey; Square Dance Katy; Father Makes Good; Joe Palooka in Humphrey Takes a Chance; Sideshow; Triple Trouble; Big Timber
1951: Casa Manana
1952: Jack and the Beanstalk; Lost in Alaska
1955: Night Freight
1956: Crashing Las Vegas; Yacqui Drums; The Women of Pitcairn Island; Hot Shots
1957: Footsteps in the Night
1961: Saintly Sinners
1967: Hillbillys in a Haunted House

NEIL YAREMA

1973: A Taste of Hell (w/ Basil Bradbury)

HAL YATES

1948: Variety Time (compilations)
1951: Footlight Varieties (compilations)

LEROY YATES

1974: Forgotten Island of Santosha (doc.)

PETER YATES (1929-) Brit.

(American films only)
1968: Bullitt
1969: John and Mary
1972: The Hot Rock
1973: The Friends of Eddie Coyle
1974: For Pete's Sake
1976: Mother, Jugs and Speed
1977: The Deep
1979: Breaking Away

IRVIN S. YEAWORTH, JR.

1958: The Blob
1959: 4D Man
1960: Dinosaurs!
1967: Way Out

LOREES YERBY

1972: Richard

BUD YORKIN (1926-)

1963: Come Blow Your Horn
1965: Never Too Late
1967: Divorce American Style
1968: Inspector Clouseau
1970: Start the Revolution Without Me
1973: The Thief Who Came to Dinner

ANTHONY YOUNG

1931: Heroes All (doc.); The Trail of the Caravan (doc.); In the Land of the Vikings (doc.); The Cradle of Faith (doc.)

GAY ALEXANDER YOUNG

1974: End of August

HAROLD YOUNG (1897-)

1934: The Scarlet Pimpernel (Brit.); Too Many Millions
1935: Without Regret
1936: Woman Trap; My American Wife
1937: Let Them Live; 52nd Street
1938: The Storm; Little Tough Guy
1939: The Forgotten Woman; Newsboys' Home; Sabotage; Hero for a Day; Code of the Streets
1940: Dreaming Out Loud
1941: Bachelor Daddy; Swing It, Soldier
1942: The Mummy's Tomb; There's One Born Every Minute; Juke Box Jennie; Robber Racketeers
1943: Hi Ya Chum; I Escaped from the Gestapo; Hi Buddy; Spy Train
1944: Machine Gun Mama
1945: The Frozen Ghost; I'll Remember April; Jungle Captive; Song of the Sarong
1948: Citizen Saint

JEFFREY YOUNG

1971: Been Down So Long It Looks Like Up to Me

MALCOLM YOUNG

1962: Trauma

ROBERT M. YOUNG

1956: Secrets of the Reef (doc. w/ others)
1977: Short Eyes
1979: Rich Kids; Alambrista!

TERENCE YOUNG (1915-) Brit.

(American films only)
1967: Wait Until Dark; Triple Cross
1969: The Christmas Tree
1974: The Klansmen
1979: Sidney Sheldon's Bloodline

TONY YOUNG

1959: Hidden Homicide

LARRY YUST

1973: Trick Baby
1974: Homebodies

ALFREDO ZACHARIAS

1979: The Bees

FRANK ZAPPA

1971: Two Hundred Motels
1979: Baby Snakes

FRANCO ZEFFIRELLI (1923-   ) Ital.

(American films only)
1979: The Champ

ALFRED ZEISLER

1937: Romance and Riches (Brit.)
1944: Enemy of Women
1946: Fear
1949: Parole, Inc.; Alimony

ROBERT ZEMECKIS

1978: I Wanna Hold Your Hand

WILLIAM ZENS

1974: Hot Summer in Barefoot County
1975: Truckin' Man

HOWARD ZIEFF

1973: Slither
1975: Hearts of the West

1978: House Calls
1979: The Main Event

VERNON ZIMMERMAN (c. 1940-)

1963: The College (Can.)
1972: Deadhead Miles (unreleased); The Unholy Rollers

FRED ZINNEMANN (1907-)

1935: Waves
1942: Eyes in the Night; Kid Glove Killer
1944: The Seventh Cross
1946: My Brother Talks to Horses; Little Mr. Jim
1948: Act of Violence; The Search
1950: The Men
1951: Teresa
1952: High Noon; The Member of the Wedding
1953: From Here to Eternity
1955: Oklahoma!
1957: A Hatful of Rain
1959: The Nun's Story
1960: The Sundowners
1964: Behold a Pale Horse
1966: A Man for All Seasons (Brit.)
1973: The Day of the Jackal
1977: Julia

JOSEPH ZITO

1975: Abduction

ALBERT ZUGSMITH (1910-)

1960: College Confidential; Sex Kittens Go to College; The Private Lives of Adam and Eve (w/ Mickey Rooney)
1961: Dondi
1962: Confessions of an Opium Eater
1966: On Her Bed of Roses
1967: Movie Star, American Style, or, LSD, I Hate You!

FRANK ZUNIGA

1978: The Further Adventures of the Wilderness Family Part II

MARTIN ZWEIBACK

1971: Cactus in the Snow